*FM 21-26

FIELD MANUAL
No. 21-26

HEADQUARTERS
DEPARTMENT OF THE ARMY
Washington, DC, 7 May 1993

MAP READING AND LAND NAVAGATION

CONTENTS

Page

DISTRIBUTION RESTRICTION Approved for public release; distribution is unlimited.

*This publication supersedes FM 2-26, 30 September 1987.

PREFACE

The purpose of this field manual is to provide a standardized source document for Armywide reference on map reading and land navigation. It applies to every soldier in the Army regardless of service branch, MOS, or rank. This manual contains both doctrine and training guidance on these subjects. Part One addresses map reading and Part Two, land navigation. The appendixes include a list of exportable training materials, a matrix of land navigation tasks, an introduction to orienteering, and a discussion of several devices that can assist the soldier in land navigation.

The proponent of this publication is the US Army Infantry School. Send comments and recommendations on DA Form 2028 directly to Commandant, US Army Infantry School, ATTN: ATSH-IN-S3, FortBenning, GA 31905-5596.

Unless this publication states otherwise, masculine nouns and pronouns do not refer exclusively to men.

PART ONE
MAP READING

CHAPTER 1
TRAINING STRATEGY

This manual is in to an Armywide need for a new map reading and land navigation training strategy based on updated doctrine. This chapter describes and illustrates this approach to teaching these skills.

1-1. THE BUILDING-BLOCK APPROACH

Institution courses are designed to prepare the soldier for a more advanced duty position in his unit. The critical soldiering skills of move, shoot, and communicate must be trained, practiced, and sustained at every level in the schools as well as in the unit. The map reading and land navigation skills taught at each level are critical to the soldiering skills of the duty position for which he is being school-trained. Therefore, they are also a prerequisite for a critical skill at a more advanced level.

a. A soldier completing initial entry training must be prepared to become a team member. He must be proficient in the basic map reading and dead reckoning skills.

b. After completing the primary leadership development course, a soldier should be ready to be a team leader. This duty position requires expertise in the skills of map reading, dead reckoning, and terrain association.

c. A soldier completing the basic NCO course has been trained for the squad leader position. Map reading and land navigation at skill level 3 requires development of problem-solving skills; for example, route selection and squad tactical movement.

d. At skill level 4, the soldier completing the advanced NCO course is prepared to assume the duty position of platoon sergeant or operations NCO. Planning tactical movements, developing unit sustainment, and making decisions are the important land navigation skills at this level.

e. Officers follow similar progression. A new second lieutenant must have mastered map reading and land navigation skills, and have an aptitude for dead reckoning and terrain association.

(1) After completing a branch-specific officer basic course, the officer must be prepared to assume the duties and responsibilities of a platoon leader. He will be required to execute the orders and operations of his commander. Map reading and land navigation at thi level require development of the problem-solving skill of route selection and tactical movement.

(2) After completing the officer advanced course the officer is prepared to assume the duties an responsibilities of a company commander or primar staff officer. The commander must plan and execut operations with full consideration to all aspects c navigation. The staff officer must recommen battlefield placement of all administrative, logistica and personnel resources. These recommendation cannot be tactically sound unless the estimate proces includes a detailed analysis of the area of operation This ability requires expertise in all map reading an navigation skills to include the use of nonmilitary map aerial photographs, and terrain analysis with respect t both friendly and enemy forces. The commander/staf officer must plan and execute a program to develop th unit's Train the Trainer Program for land navigation.

f. A program of demonstrated proficiency of a the preceding skill levels to the specified conditions an standards is a prerequisite to the successf implementation of a building-block training approac This approach will reflect duty position responsibilitie in map reading and land navigation. An understandin of the fundamental techniques of dead reckoning c field-expedient methods is a basic survival skill tha each soldier must develop at the initial entry level. Thi provides a support foundation for more interpretiv analysis at intermediate skill levels 2 and 3, with fina progression to level 4. Mastery of all map reading an land navigation tasks required in previous duty position is essential for the sequential development c increasingly difficult abilities. This building-bloc approach is supported by scope statements. It is part c the training doctrine at each level in the institutiona training environment of each course.

g. Exportable training end instructor support/cerication packages, based upon the updated map reading d land navigation field manual, are being developed. novative training devices and materials are being veloped for use in the institution, ROTC regions, and e field. (See Appendixes E and H.)

2. ARMYWIDE IMPLEMENTATION

mandatory core of critical map reading and land navition tasks and a list of electives will be provided to ch TRADOC service school and FORSCOM ofessional development school. Standardization will be hieved through the mandatory core. Exportable training material will be made available to support Armywide implementation.

1-3. SAFETY

Plan to brief and enforce all safety regulations established by local range control. Coordinate the mode of evacuation of casualties through the appropriate channels. Review all installation safety regulations. Unit leaders must complete a thorough terrain reconnaissance before using an area for land navigation training. They should look for dangerous terrain, heavy trafficked roads, water obstacles, wildlife, and training debris.

MAPS

Cartography is the art and science of expressing the known physical features of the earth graphical by maps and charts. No one knows who drew, molded, laced together, or scratched out us the dirt the first map. But a study of history reveals that the most pressing demands for accuracy and detail in mapping have come as the result of military needs. Today, the complexities of tactical operations and deployment of troops is such that it is essential for all soldiers to be able to read and interpret their maps in order to move quickly and effectively on the battlefield. This chapter explains maps; it includes the definition and purpose of a map and describes map security, types, categories, and scales.

2-1. DEFINITION

A map is a graphic representation of a portion of the earth's surface drawn to scale, as seen from above. It uses colors, symbols, and labels to represent features found on the ground. The ideal representation would be realized if every feature of the area being mapped could be shown in true shape. Obviously this is impossible, and an attempt to plot each feature true to scale would result in a product impossible to read even with the aid of a magnifying glass.

a. Therefore, to be understandable, features must be represented by conventional signs and symbols. To be legible, many of these must be exaggerated in size, often far beyond the actual ground limits of the feature represented. On a 1:250,000 scale map, the prescribed symbol for a building covers an area about 500 feet square on the ground; a road symbol is equivalent to a road about 520 feet wide on the ground; the symbol for a single-track railroad (the length of a cross-tie) is equivalent to a railroad cross-tie about 1,000 feet on the ground.

b. The portrayal of many features requires similar exaggeration. Therefore, both the selection of features to be shown, as well as their portrayal, are in accord with the guidance established by the Defense Mapping Agency.

2-2. PURPOSE

A map provides information on the existence, the location of, and the distance between ground features, such as populated places and routes of travel and communication. It also indicates variations in terrain, heights of natural features, and the extent of vegetation cover. With our military forces dispersed throughout the world, it is necessary to rely on maps to provide information to our combat elements and to resolve logistical operations far from our shores. Soldiers and materials must be transported, stored, and placed into operation at the proper time and place. Much of this planning must be done by using maps. Therefore, any operation requires supply of maps; however, the finest maps available a worthless unless the map user knows how to read them.

2-3. PROCUREMENT

Most military units are authorized a basic load of map Local command supplements to AR 115-11 provi tables of initial allowances for maps. Map requisition and distribution follow the channels of Defense Mappi Agency Hydrographic, Topographic Center's Office Distribution and Services. In the division, howeve maps are a responsibility of the G2 section.

a. To order a map, refer to the DMA catalc located at your S2/G2 shop. Part 3 of this catalc Topographic Maps, has five volumes. Using the delineated map index, find the map or maps you wa based upon the location of the nearest city. With th information, order maps using the following forms:

(1) *Standard Form 344.* It can be typed handwritten; it is used for mailing or over-the-count service.

(2) *Department of Defense Form 1348.* Same as S 344. You can order copies of only one map sheet on ea form.

(3) *Department of Defense Form 1348M.* This is punch card form for AUDODIN ordering.

(4) *Department of Defense Form 173.* This is message form to be used for urgent ordering.

With the exception of the message form (DD 173), t numbered sections of all forms are the same. F example:. In block 1, if you are in CONUS, ent "AOD;" if you are overseas, enter "AO4." In block 2, u one of the following codes for your location.

Location	Code
Europe	CS7
Hawaii	HM9
Korea	WM4
Alaska	WC1
Panama	HMJ
CONUS	HM8

your supply section will help you fill out the rest of the form.

b. Stock numbers are also listed in map catalogs, which are available at division and higher levels and occasionally in smaller units. A map catalog consists of small-scale maps upon which the outlines of the individual map sheets of a series have been delineated. Another document that is an aid to the map user is the gazetteer. A gazetteer lists all the names appearing on a map series of a geographical area, a designation that identifies what is located at that place name, a grid reference, a sheet number of the map upon which the name appeared, and the latitude and longitude of the named features. Gazetteers are prepared for maps of foreign areas only.

2-4. SECURITY

All maps should be considered as documents that require special handling. If a map falls into unauthorized hands, it could easily endanger military operations by providing information of friendly plans or areas of interest to the enemy. Even more important would be a map on which the movements or positions of friendly soldiers were marked. It is possible, even though the markings on a map have been erased, to determine some of the information that had been marked upon it. **Maps are documents that must not fall into unauthorized hands.**

a. If a map is no longer needed, it must be turned in to the proper authority. If a map is in danger of being captured, it must be destroyed. The best method of destruction is by burning it and scattering the ashes. If burning is not possible, the map can be torn into small pieces and scattered over a wide area.

b. Maps of some areas of the world are subject to third party limitations. These are agreements that permit the United States to make and use maps of another country provided these maps are not released to any third party without permission of the country concerned. Such maps require special handling.

c. Some maps may be classified and must be handled and cared for in accordance with AR 380-5 and, if applicable, other local security directives.

2-5. CARE

Maps are documents printed on paper and require protection from water, mud, and tearing. Whenever possible, a map should be carried in a waterproof case, in a pocket, or in some other place where it is handy for use but still protected.

a. Care must also be taken when using a map since it may have to last a long time. If it becomes necessary to mark a map, the use of a pencil is recommended. Use light lines so they may be erased easily without smearing and smudging, or leaving marks that may cause confusion later. If the map margins must be trimmed for any reason, it is essential to note any marginal information that may be needed later, such as grid data and magnetic declination.

b. Special care should be taken of a map that is being used in a tactical mission, especially in small units; the mission may depend on that map. All members of such units should be familiar with the map's location at all times.

c. Appendix B shows two ways of folding a map.

2-6. CATEGORIES

The DMA's mission is to provide mapping, charting, and all geodesy support to the armed forces and all other national security operations. DMA produces four categories of products and services: hydrographic, topographic, aeronautical, and missile and targeting. Military maps are categorized by scale and type.

a. **Scale.** Because a map is a graphic representation of a portion of the earth's surface drawn to scale as seen from above, it is important to know what mathematical scale has been used. You must know this to determine ground distances between objects or locations on the map, the size of the area covered, and how the scale may affect the amount of detail being shown. The mathematical scale of a map is the ratio or fraction between the distance on a map and the corresponding distance on the surface of the earth. Scale is reported as a representative fraction (RF) with the map distance as the numerator and the ground distance as the denominator.

$$Representative\ fraction\ (scale\ ) = \frac{map\ distance}{ground\ distance}$$

As the denominator of the RF gets larger and the ratio gets smaller, the scale of the map decreases. Defense Mapping Agency maps are classified by scale into three categories. They are small-, medium-, and large-scale maps (Figure 2-1). The terms **"small scale," "medium scale,"** and **"large scale"** may be confusing when read in conjunction with the number. However, if the number is viewed as a fraction, it quickly becomes apparent that 1:600,000 of something is smaller than 1:75,000 of the same thing. Therefore, the larger the number after 1:, the smaller the scale of the map.

(1) *Small.* Those maps with scales of 1:1,000,000 and smaller are used for general planning and for strategic studies (bottom map in Figure 2-1). The standard small-

scale map is 1:1,000,000. This map covers a very large land area at the expense of detail.

(2) *Medium.* Those maps with scales larger than 1:1,000,000 but smaller than 1:75,000 are used for operational planning (center map in Figure 2-1). They contain a moderate amount of detail, but terrain analysis is best done with the large-scale maps described below. The standard medium-scale map is 1:250,000. Medium scale maps of 1:100,000 are also frequently encountered.

(3) *Large.* Those maps with scales of 1:75,000 and larger are used for tactical, administrative, and logistical planning (top map in Figure 2-1). These are the maps that you as a soldier or junior leader are most likely to encounter. The standard large-scale map is 1:50,000; however, many areas have been mapped at a scale of 1:25,000.

Figure 2-1. Scale classifications.

b. **Types.** The map of choice for land navigators is the 1:50,000 scale military topographic map. It is important, however, that you know how to use the many other products available from the DMA as well. When operating in foreign places, you may discover that DMA map products have not yet been produced to cover your particular area of operations, or they may not be available to your unit when you require them. Therefore, you must be prepared to use maps produced by foreign governments that may or may not meet the standards for accuracy set by DMA. These maps often use symbols that resemble those found on DMA maps but which have completed different meanings. There may be other times when you must operate with the only map you can obtain. This might be a commercially produced map run off on a copy machine at higher headquarters. In Grenada, many of our troops used a British tourist map.

(1) *Planimetric map.* This is a map that presents only the horizontal positions for the features represented. It is distinguished from a topographic map by the omission of relief, normally represented by contour lines. Some times, it is called a line map.

(2) *Topographical map.* This is a map that portrays terrain features in a measurable way (usually through use of contour lines), as well as the horizontal positions of the features represented. The vertical positions, or relief, are normally represented by contour lines on military topographic maps. On maps showing relief, the elevations and contours are measured from specific vertical datum plane, usually mean sea level. Figure 3-1 shows a typical topographic map.

(3) *Photomap.* This is a reproduction of an aerial photograph upon which grid lines, marginal data, place names, route numbers, important elevations, boundaries, and approximate scale and direction have been added. (See Chapter 8)

(4) *Joint operations graphics.* These are based on the format of standard 1:250,000 medium-scale military topographic maps, but they contain additional information needed in joint air-ground operations (Figure 2-2). Along the north and east edges of the graphic, detail is extended beyond the standard map sheet to provide overlap with adjacent sheets. These maps are produced both in ground and air formats. Each version is identified in the lower margin as either Joint Operations Graphic (Air)

Figure 2-2. Joint operations graphic (air).

r Joint Operations Graphic (Ground). The topographic
formation is identical on both, but the ground version
nows elevations and contour in meters and the air
ersion shows them in feet. Layer (elevation) tinting and
elief shading are added as an aid to interpolating relief.
oth versions emphasize airlanding facilities (shown in
urple), but the air version has additional symbols to
lentify aids and obstructions to air navigation. (See
ppendix D for additional information.)

(5) *Photomosaic*. This is an assembly of aerial
hotographs that is commonly called a mosaic in
opographic usage. Mosaics are useful when time does
ot permit the compilation of a more accurate map. The
ccuracy of a mosaic depends on the method employed
a its preparation and may vary from simply a good
ictorial effect of the ground to that of a planimetric
tap.

(6) *Terrain model*. This is a scale model of the
rrain showing features, and in large-scale models
nowing industrial and cultural shapes. It provides a
teans for visualizing the terrain for planning or
tdoctrination purposes and for briefing on assault
tndings.

(7) *Military city map*. This is a topographic map
tsually at 1:12,550 scale, sometimes up to 1:5,000),
nowing the details of a city. It delineates streets and
nows street names, important buildings, and other
lements of the urban landscape important to navigation
nd military operations in urban terrain. The scale of a
tilitary city map depends on the importance and size of
te city, density of detail, and available intelligence
tformation.

(8) *Special maps* These are maps for special
urposes, such as trafficability, communications, and
ssault maps. They are usually in the form of an
verprint in the scales smaller than 1:100,000 but larger
tan 1:1,000,000. A special purpose map is one that has
een designed or modified to give information not
overed on a standard map. The wide range of subjects
tat could be covered under the heading of special
urpose maps prohibits, within the scope of this manual,
tore than a brief mention of a few important ones.
ome of the subjects covered are:

- Terrain features.
- Drainage characteristics.
- Vegetation.
- Climate.
- Coasts and landing beaches.
- Roads and bridges.
- Railroads.
- Airfields.
- Urban areas.
- Electric power.
- Fuels.
- Surface water resources.
- Ground water resources.

- Natural construction materials.
- Cross-country movements.
- Suitability for airfield construction.
- Airborne operations.

2 7. MILITARY MAP SUBSTITUTES

If military maps are not available, substitutes will have
to be used These can range from foreign military or
commercial maps to field sketches. The DMA can
provide black and white reproductions of many foreign
maps and can produce its own maps based upon
intelligence

a. **Foreign Maps.** These are maps that have been
compiled by nations other than our own. When these
must be used, the marginal information and grids are
changed to conform to our standards if time permits.
The scales may differ from our maps, but they do express
the ratio of map distant to ground distance and can be
used in the same way. The legend must be used since
the map symbols almost always differ from ours.
Because the accuracy of foreign maps varies
considerably, they are usually evaluated in regard to
established accuracy standards before they are issued to
our troops. (See Appendix K for additional
information.)

b. **Atlases.** Atlases are collections of maps of
regions, countries, continents, or the world. Such maps
are accurate only to a degree and can be used for general
information only.

c. **Geographic Maps.** These maps give an overall
idea of the mapped area in relation to climate,
population, relief, vegetation, and hydrography. They
also show general location of major urban areas.

d. **Tourist Road Maps.** These are maps of a
region in which the main means of transportation and
areas of interest are shown. Some of these maps show
secondary networks of roads, historic sites, museums,
and beaches in detail. They may contain road and time
distance between points. Careful consideration should
be exercised about the scale when using these maps.

e. **City/Utility Maps.** These are maps of urban
areas showing streets, water ducts, electricity and
telephone lines, and sewers.

f. **Field Sketches**. These are preliminary
drawings of an area or piece of terrain. (See Appendix
A)

g. **Aerial Photographs.** These can be used as map
supplements or substitutes to help you analyze the
terrain, plan your route, or guide your movement. (See
Chapter 8 for additional information).

2-8. STANDARDS OF ACCURACY

Accurate, is the degree of conformity with which
horizontal positions and vertical values are represented
on a map in relation to an established standard. This
standard is determined by the DMA based on user
requirements. A map can be considered to meet
accuracy requirement standards unless otherwise
specified in the marginal information.

CHAPTER 3

MARGINAL INFORMATION AND SYMBOLS

A map could be compared to any piece of equipment, in that before it is placed into operation the user must read the instructions. It is important that you, as a soldier, know how to read these instructions. The most logical place to begin is the marginal information and symbols, where useful information telling about the map is located and explained. All maps are not the same, so it becomes necessary every time a different map is used to examine the marginal information carefully.

3-1. MARGINAL INFORMATION ON A MILITARY MAP

Figure 3-1, page 3-3, shows a reduced version of a large-scale topographic map. The circled numbers indicate the items of marginal information that the map user needs to know. These circled numbers correspond to the following listed items.

a. **Sheet Name (1).** The sheet name is found in bold print at the center of the top and in the lower left area of the map margin. A map is generally named for the largest settlement contained within the area covered by the sheet, or for the largest natural feature located entirely within the area at the time the map was drawn.

b. **Sheet Number (2).** The sheet number is found in bold print in both the upper right and lower left areas of the margin, and in the center box of the adjoining sheets diagram, which is found in the lower right margin. It is used as a reference number to link specific maps to overlays, operations orders, and plans. For maps at 1:100,000 scale and larger, sheet numbers are based on an arbitrary system that makes possible the ready orientation of maps at scales of 1:100,000, 1:50,000, and 1:25,000.

c. **Series Name (3).** The map series name is found in the same bold print as the sheet number in the upper left corner of the margin. The name given to the series is generally that of a major political subdivision, such as a state within the United States or a European nation. A map series usually includes a group of similar maps at the same scale and on the same sheet lines or format designed to cover a particular geographic area. It may also be a group of maps that serve a common purpose, such as the military city maps.

d. **Scale (4).** The scale is found both in the upper left margin after the series name, and in the center of the lower margin. The scale note is a representative fraction that gives the ratio of a map distance to the corresponding distance on the earth's surface. For example, the scale note 1:50,000 indicates that one unit of measure on the map equals 50,000 units of the same measure on the ground.

e. **Series Number (5).** The series number is found in both the upper right margin and the lower left margin. It is a sequence reference expressed either as a four-digit numeral (1125) or as a letter, followed by a three- or four-digit numeral (M661; T7110).

f. **Edition Number (6).** The edition number is found in bold print in the upper right area of the top margin and the lower left area of the bottom margin. Editions are numbered consecutively; therefore, if you have more than one edition, the highest numbered sheet is the most recent. Most military maps are now published by the DMA, but older editions of maps may have been produced by the US Army Map Service. Still others may have been drawn, at least in part, by the US Army Corps of Engineers, the US Geological Survey, or other agencies affiliated or not with the United States or allied governments. The credit line, telling who produced the map, is just above the legend. The map information date is found immediately below the word "LEGEND" in the lower left margin of the map. This date is important when determining how accurately the map data might be expected to match what you will encounter on the ground.

g. **Index to Boundaries (7).** The index to boundaries diagram appears in the lower or right margin of all sheets. This diagram, which is a miniature of the map, shoes the boundaries that occur within the map area, such as county lines and state boundaries.

h. **Adjoining Sheets Diagram (8).** Maps at all standard scales contain a diagram that illustrates the adjoining sheets. On maps at 1:100,000 and larger scales and at 1:1,000,000 scale, the diagram is called the index to adjoining sheets. It consists of as many rectangles representing adjoining sheets as are necessary to surround the rectangle that represents the sheet under consideration. The diagram usually contains nine rectangles, but the number may vary depending on the cations of the adjoining sheets. All represented sheets are identified by their sheet numbers. Sheets of an adjoining series, whether published or planned, that are the same scale are represented by dashed lines. The series number of the adjoining series is indicated along the appropriate side of the division line between the series.

i. **Elevation Guide (9).** This is normally found in the lower right margin. It is a miniature characterization of the terrain shown. The terrain is represented by bands of elevation, spot elevations, and major drainage features. The elevation guide provides the map reader with a means of rapid recognition of major landforms.

j. **Declination Diagram (10).** This is located in the lower margin of large-scale maps and indicates the angular relationships of true north, grid north, and magnetic north. On maps at 1:250,000 scale, this information is expressed as a note in the lower margin. In recent edition maps, there is a note indicating the conversion of azimuths from grid to magnetic and from magnetic to grid next to the declination diagram.

k. **Bar Scales (11).** These are located in the center of the lower margin. They are rulers used to convert map distance to ground distance. Maps have three or more bar scales, each in a different unit of measure. Care should be exercised when using the scales, especially in the selection of the unit of measure that is needed.

l. **Contour Interval Note (12).** This note is found in the center of the lower margin normally below the bar scales. It states the vertical distance between adjacent contour lines of the map. When supplementary contours are used, the interval is indicated. In recent edition maps, the contour interval is given in meters instead of feet.

m. **Spheroid Note (13).** This note is located in the center of the lower margin. Spheriods (ellipsoids) have specific parameters that define the X Y Z axes of the earth. The spheriod is an integral part of the datum.

n. **Grid Note (14).** This note is located in the center of the lower margin. It gives information pertaining to the grid system used and the interval between grid lines, and it identifies the UTM grid zone number.

o. **Projection Note (15).** The projection system is the framework of the map. For military maps, this framework is of the conformal type; that is, small areas on the surface of the earth retain their true shapes on the projection; measured angles closely approximate true values; and the scale factor is the same in all directions from a point. The projection note is located in the center of the lower margin. Refer to DMA for the development characteristics of the conformal-type projection systems.

(1) Between 80° south and 84° north, maps at scales larger than 1:500,000 are based on the transverse mercator projection. The note reads TRANSVERSE MERCATOR PROJECTION.

(2) Between 80° south and 84° north, maps at 1:1,000,000 scale and smaller are based on standard parallels of the lambert conformal conic projection. The note reads, for example, LAMBERT CONFORMAL CONIC PROJECTIONS 36° 40' N AND 39° 20' N.

(3) Maps of the polar regions (south of 80° south and north of 84 north) at 1:1,000,000 and larger scales are based on the polar stereographic projection. The note reads POLAR STEREOGRAPHIC PROJECTION.

p. **Vertical Datum Note (16).** This note is located in the center of the lower margin. The vertical datum or vertical-control datum is defined as any level surface (for example, mean sea level) taken as a surface of reference from which to reckon elevations. In the United States, Canada, and Europe, the vertical datum refers to the mean sea level surface. However, in parts of Asia and Africa, the vertical-control datum may vary locally and is based on an assumed elevation that has no connection to any sea level surface. Map readers should habitually check the vertical datum note on maps, particularly if the map is used for low-level aircraft navigation, naval gunfire support, or missile target acquisition.

q. **Horizontal Datum Note (17).** This note is located in the center of the lower margin. The horizontal datum or horizontal-control datum is defined as a geodetic reference point (of which five quantities are known: latitude, longitude, azimuth of a line from this point, and two constants, which are the parameters of reference ellipsoid). These are the basis for horizontal-control surveys. The horizontal-control datum may extend over a continent or be limited to a small local area. Maps and charts produced by DMA are produced on 32 different horizontal-control data. Map readers should habitually check the horizontal datum note on every map or chart, especially adjacent map sheets. This is to ensure the products are based on the same horizontal datum. If products are based on different horizontal-control data, coordinate transformations to a common datum must be performed. UTM coordinates from the same point computed on different data may differ as much as 900 meters.

r. **Control Note (18).** This note is located in the center of the lower margin. It indicates the special agencies involved in the control of the technical aspects of all the information that is disseminated on the map.

s. **Preparation Note (19).** This note is located in the center of the lower margin. It indicates the agency responsible for preparing the map.

t. **Printing Note (20).** This note is also located in the center of the lower margin. It indicates the agency responsible for printing the map and the date the map was printed. The printing data should not be used to determine when the map information was obtained.

Figure 3-1. Topographic map.

u. **Grid Reference Box (21).** This box is normally located in the center of the lower margin. It contains instructions for composing a grid reference.

v. **Unit Imprint and Symbol (22).** The unit imprint and symbol is on the left side of the lower margin. It identifies the agency that prepared and printed the map with its respective symbol. This information is important to the map user in evaluating the reliability of the map.

w. **Legend (23).** The legend is located in the lower left margin. It illustrates and identifies the topographic symbols used to depict some of the more prominent features on the map. The symbols are not always the same on every map. Always refer to the legend to avoid errors when reading a map.

-2. ADDITIONAL NOTES

Not all maps contain the same items of marginal information. Under certain conditions, special notes and scales may be added to aid the map user. The following are examples:

a. **Glossary.** This is an explanation of technical terms or a translation of terms on maps of foreign areas where the native language is other than English.

b. **Classification.** Certain maps require a note indicating the security classification. This is shown in the upper and lower margins.

c. **Protractor Scale.** This scale may appear in the upper margin on some maps. It is used to lay out the magnetic-grid declination for the map which, in turn, is used to orient the map sheet with the aid of the lensatic compass.

d. **Coverage Diagram.** On maps at scales of 1:100,000 and larger, a coverage diagram may be used. It is normally in the lower or right margin and indicates the methods by which the map was made, dates of photography, and reliability of the sources. On maps at 1:250,000 scale, the coverage diagram is replaced by a reliability diagram.

e. **Special Notes (24).** A special note is any statement of general information that relates to the mapped area. It is normally found in the lower right margin. For example: This map is red-light readable.

f. **User's Note (25).** This note is normally located in the lower right-hand margin. It requests cooperation in correcting errors or omissions on the map. Errors should be marked and the map forwarded to the agency identified in the note.

g. **Stock Number Identification (26).** All maps published by the DMA that are in the Department of the Army map supply system contain stock number identifications that are used in requisitioning map supplies. The identification consists of the words "STOCK NO" followed by a unique designation that is composed of the series number, the sheet number of the individual map and, on recently printed sheets, the edition number. The designation is limited to 15 units (letters and numbers). The first 5 units are allotted to the series number; when the series number is less than 5 units, the letter "X" is substituted as the fifth unit. The sheet number is the next component; however, Roman numerals, which are part of the sheet number, are converted to Arabic numerals in the stock number. The last 2 units are the edition number; the first digit of the edition number is a zero if the number is less than 10. If the current edition number is unknown, the number 01 is used. The latest available edition will be furnished. Asterisks are placed between the sheet number and the edition number when necessary to ensure there are at least 11 units in the stock number.

h. **Conversion Graph (27).** Normally found in the right margin, this graph indicates the conversion of different units of measure used on the map.

3-3. TOPOGRAPHIC MAP SYMBOLS

The purpose of a map is to permit one to visualize an area of the earth's surface with pertinent features properly positioned. The map's legend contains the symbols most commonly used in a particular series or on that specific topographic map sheet. Therefore, the legend should be referred to each time a new map is used. Every effort is made to design standard symbols that resemble the features they represent. If this is not possible, symbols are selected that logically imply the features they portray. For example, an open-pit mining operation is represented by a small black drawing of a crossed hammer and pickax.

a. Ideally, all the features within an area would appear on a map in their true proportion, position, and shape. This, however, is not practical because many of the features would be unimportant and others would be unrecognizable because of their reduction in size.

b. The mapmaker has been forced to use symbols to represent the natural and man-made features of the earth's surface. These symbols resemble, as closely as possible, the actual features themselves as viewed from above. They are positioned in such a manner that the center of the symbol remains in its true location. An exception to this would be the position of a feature adjacent to a major road. If the width of the road has been exaggerated, then the feature is moved from its true position to preserve its relation to the road. Field Manual 21-31 gives a description of topographic features and abbreviations authorized for use on our military maps.

3-4. MILITARY SYMBOLS

In addition to the topographic symbols used to represent the natural and man-made features of the earth, military personnel require some method for showing identity, size, location, or movement of soldiers; and military activities and installations. The symbols used to represent these military features are known as military symbols. These symbols are not normally printed on maps because the features and units that they represent are constantly moving or changing; military security is also a consideration. They do appear in special maps and overlays (Chapter 7). The map user draws them in, in accordance with proper security precautions. Refer to FM 101-5-1 for complete information on military symbols.

3-5.   COLORS USED ON A MILITARY MAP
By the fifteenth century, most European maps were carefully colored. Profile drawings of mountains and hills were shown in brown, rivers and lakes in blue, vegetation in green, roads in yellow, and special information in red. A look at the legend of a modern map confirms that the use of colors has not changed much over the past several hundred years. To facilitate the identification of features on a map, the topographical and cultural information is usually printed in different colors. These colors may vary from map to map. On standard large-scale topographic map, the colors use and the features each represent are:

a.   **Black.** Indicates cultural (man-made) feature such as buildings and roads, surveyed spot elevation and all labels.

b.   **Red-Brown.** The colors red and brown a combined to identify cultural features, all relief feature non-surveyed spot elevations, and elevation, such a contour lines on red-light readable maps.

c.   **Blue.** Identifies hydrography or water feature such as lakes, swamps, rivers, and drainage.

d.   **Green.** Identifies vegetation with militar significance, such as woods, orchards, and vineyards.

e.   **Brown.** Identifies all relief features ar elevation, such as contours on older edition maps, ar cultivated land on red-light readable maps.

f.   **Red.** Classifies cultural features, such a populated areas, main roads, and boundaries, on old maps.

g.   **Other.** Occasionally other colors may be use to show special information. These are indicated in th marginal information as a rule.

GRIDS

This chapter covers how to determine and report positions on the ground in terms of their locations on a map. Knowing where you are (position fixing) and being able to communicate that knowledge is crucial to successful land navigation as well as to the effective employment of direct and indirect fire, tactical air support, and medical evacuation. It is essential for valid target acquisition; accurate reporting of nuclear, biological, and chemical (NBC) contamination and various danger areas; and obtaining emergency resupply. Few factors contribute as much to the survivability of troops and equipment and to the successful accomplishment of a mission as always knowing where you are. The chapter includes explanations of geographical coordinates, Universal Transverse Mercator grids, the military grid reference system, and the use of grid coordinates.

4-1. REFERENCE SYSTEM
In a city, it is quite simple to find a location; the streets are named and the buildings have numbers. The only thing needed is the address. However, finding locations in undeveloped areas or in unfamiliar parts of the world can be a problem. To cope with this problem, a uniform and precise system of referencing has been developed.

4-2. GEOGRAPHIC COORDINATES
One of the oldest systematic methods of location is based upon the geographic coordinate system. By drawing a set of east-west rings around the globe (parallel to the equator), and a set of north-south rings crossing the equator at right angles and converging at the poles, a network of reference lines is formed from which any point on the earth's surface can be located.

a. The distance of a point north or south of the equator is known as its latitude. The rings around the earth parallel to the equator are called parallels of latitude or simply parallels. Lines of latitude run east-west but north-south distances are measured between them.

b. A second set of rings around the globe at right angles to lines of latitude and passing through the poles are known as meridians of longitude or simply meridians. One meridian is designated as the prime meridian. The prime meridian of the system we use runs through Greenwich, England and is known as the Greenwich meridian. The distance east or west of a prime meridian to a point is known as its longitude. Lines of longitude (meridians) run north-south but east-west distances are measured between them (Figures 4-1 and 4-2).

c. Geographic coordinates are expressed in angular measurement. Each circle is divided into 360

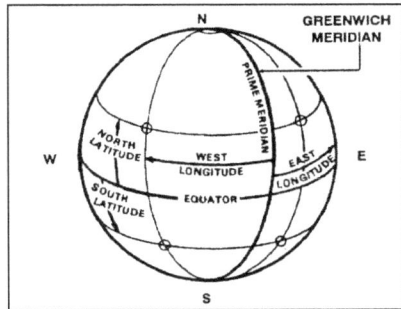

Figure 4-1. Prime meridian and equator.

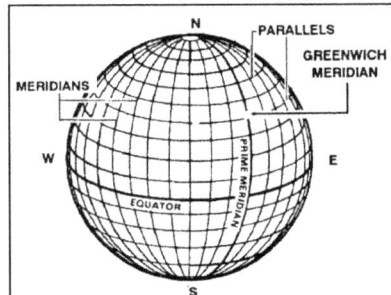

Figure 4-2. Reference lines.

degrees each degree into 60 minutes, and each minute into 60 seconds. The degree is symbolized by °, the minute by ', and the second by ". Starting with 0° at the equator, the parallels of latitude are numbered to 90° both north and south. The extremities are the north pole at 90° north latitude and the south pole at 90° south latitude- Latitude can have the same numerical value north or south of the equator, so the direction N or S must always be given. Starting with 0° at the prime meridian, longitude is measured both east and west around the world. Lines east of the prime meridian are numbered to 180° and identified as east longitude; lines west of the prime meridian are numbered to 180° and identified as west longitude. The direction E or W must always he given. The line directly opposite the prime meridian, 180°, may be referred to as either east or west longitude. The values of geographic coordinates, being in units of angular measure, will mean more if they are compared with units of measure with which we are more familiar. At any point on the earth, the ground distance covered by one degree of latitude is about 111 kilometers (69 miles); one second is equal to about 30 meters (100 feet). The ground distance covered by one degree of longitude at the equator is also about 111 kilometers, but decreases as one moves north or south, until it becomes zero at the poles. For example, one second of longitude represents about 30 meters (100 feet) at the equator; but at the latitude of Washington, DC, one second of longitude is approximately 24 meters (78 feet). Latitude and longitude are illustrated in Figure 4-3.

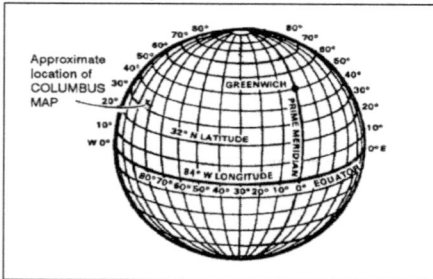

Figure 4-3. Latitude and longitude.

d. Geographic coordinates appear on all standard military maps; on some they may be the only method of locating and referencing the location of a point. The four lines that enclose the body of the map (neatlines) are latitude and longitude lines. Their values are given in degrees and minutes at each of the four corners. On a portion of the Columbus map (Figure 4-4), the figures

32°15' and 84°45' appear at the lower right corner. The bottom line of this map is latitude 32°15'00"N, and the line running up the right side is longitude 84°45'00"W. In addition to the latitude and longitude given for the four corners, there are, at regularly spaced intervals along the sides of the map, small tick marks extending into the body of the map. Each of these tick marks is identified by its latitude or longitude value. Near the top of the right side of the map is a tick mark and the number 20'. The full value for this tick marks is 32°20'00" of latitude. At one-third and two-thirds of the distance across the map from the 20' tick mark will be found a cross tick mark (grid squares 0379 and 9679) and at the far side another 20' tick mark. By connecting the tick marks and crosses with straight lines, a 32°20'00" line of latitude can be added to the map. This procedure is also used to locate the 32°25'00" line of latitude. For lines of longitude, the same procedure is followed using the tick marks along the top and bottom edges of the map.

e. After the parallels and meridians have been drawn, the geographic interval (angular distance between two adjacent lines) must be determined. Examination of the values given at the tick marks gives the interval. For most maps of scale 1:25,000, the interval is 2'30". For the Columbus map and most maps of scale 1:50,000, it is 5'00". The geographic coordinates of a point are found by dividing the sides of the geographic square in which the point is located into the required number of equal parts. If the geographic interval is 5'00 and the location of a point is required to the nearest second, each side of the geographic square must be divided into 300 equal parts (5'00" = 300"), each of which would have a value of one second. Any scale or ruler that has 300 equal divisions and is as long as or longer than the spacing between the lines may be used.

f. The following steps will determine the geographic coordinates of Wilkinson Cemetery (northwest of the town of Cusseta) on the Columbus map.
(1) Draw the parallels and meridians on the map that enclose the area around the cemetery.
(2) Determine the values of the parallels and meridians where the point falls.
Latitude 32°15'00" and 32°20'00".
Longitude 84°45'00" and 84050100".
(3) Determine the geographic interval (5'00" = 300").

(4) Select a scale that has 300 small divisions or multiples thereof (300 divisions, one second each; 150 divisions, two seconds each; 75 divisions, four seconds each, and so forth).

(5) To determine the latitude—

(a) Place the scale with the 0 of the scale on the latitude of the lowest number value (32°15'00") and the 300 of the scale on the highest numbered line (32°20'00") (1, Figure 4-4).

(b) Keeping the 0 and 300 on the two lines, slide the scale (2, Figure 4-4) along the parallels until the Wilkinson Cemetery symbol is along the edge of the numbered scale.

(c) Read the number of seconds from the scale (3, Figure 4-4), about 246.

(d) Covert the number of seconds to minutes and seconds (246" = 4'6") and add to the value of the lower numbered line (32°15'00" + 4'06" = 32°19'06") (4, Figure 4-4).

(e) The latitude is 32°19'06", but this is not enough.

(f) The latitude 32°19'06" could be either north or south of the equator, so the letter N or S must be added

Figure 4-4. Determining latitude.

to the latitude. To determine whether it is N or S, look at the latitude values at the edge of the map and find the direction in which they become larger. If they are larger going north, use N; if they are larger going south, use S.

(g) The latitude for the cemetery is 32°19'06"N.

(6) To determine the longitude, repeat the same steps but measure between lines of longitude and use E and W. The geographic coordinates of Wilkinson Cemetery should be about 32°19'06"N and 84°47'32"W (Figure 4-5).

Figure 4-5. Determining longitude.

g. To locate a point on the Columbus map (Figure4-6) when knowing the geographic coordinates, many of the same steps are followed. To locate 32°25'28"N and 84°50'56"W, first find the geographic lines within which the point fills: latitude 32°25'00" and 32°30'0"; and longitude 84°50'00" and 84°55'00". Subtract the lower latitude/longitude from the higher latitude/longitude.

(1) Place the 0 of the scale on the 32°25'00" line and the 300 on the 32°30'00". Make a mark at the number 28 on the scale (the difference between the lower and higher latitude).

(2) Place the 0 of the scale on the 84°50'00" line end the 300 on the 84°50'55". Make a mark at the number 56 on the scale (the difference between the lower and higher longitude.

(3) Draw a vertical line from the mark at 56 and a horizontal line from the mark at 28; they will intersect at 32°25'28"N and 84°50'56"W.

Figure 4-6. Determining geographic coordinates.

h. If you do not have a scale or ruler with 300 equal divisions or a map whose interval is other than 5'00" use the proportional parts method. Following the steps will determine the geographic coordinates of horizontal control station 141.

(1) Locate horizontal control station 141 in grid square (GL0784) (Figure 4-7).

(2) Find a cross in grid square GL0388 and a tick mark in grid square GL1188 with 25'.

(3) Find another cross in grid square GL0379 and a tick mark in grid square GL1179 with 20'.

(4) Enclose the control station by connecting the crosses and tick marks. The control station is between 20' and 25' (Figure 4-7).

(5) With a boxwood scale, measure the distance from the bottom line to the top line that encloses the area around the control station on the map (total distance) (Figure 4-7).

Figure 4-7. Using the proportional parts method.

(6) Measure the partial distance from the bottom line to the center of the control station (Figure 4-7). These straight-line distances are in direct proportion to the minutes and seconds of latitude and are used to set up a ratio.

(7) The total distance is 9,200 meters, and the partial distance is 5,125 meters (Figure 4-7).

(8) With the two distances and the five-minute interval converted to seconds (300"), determine the minutes and seconds of latitude using the following formula:

1. 5,125, x 300 = 1,537,500
2. 1,537,500 ÷ 9,200 = 167
3. 167 ÷ 60=2'47"
4. Add 2'47" to 32°20'00" = 32°22'47"

(9) Follow the same procedures to determine minutes and seconds of longitude (Figure 4-7).

(10) The total distance is 7,830 meters, and the partial distance is 4,000 meters (Figure 4-7).

1. 4,000 x 300 = 1,200,000
2. 1,200,000 ÷ 7,830 = 153
3. 153 ÷ 60=2'33"
4. Add 2'33" to 84°45' = 84°47'33"N

(11) The geographic coordinates of horizontal control station 141 in grid square GL0784 are 32°22'47"N latitude and 84°47'33"W longitude.

NOTE: When computing formulas, you must round off totals to the nearest whole number in step 2. In step 3, convert the fraction to seconds by multiplying the fraction by 60 and rounding off if the total is not a whole number.

i. The maps made by some nations do not have their longitude values based on the prime meridian that passes through Greenwich, England. Table 4-1 shows the prime meridians that may be used by other nations. When these maps are issued to our soldiers, a note usually appears in the marginal information giving the difference between our prime meridian and the one used on the map.

4-3. MILITARY GRIDS

An examination of the transverse Mercator projection, which is used for large-scale military maps, shows that most lines of latitude and longitude are curved lines. The quadrangles formed by the intersection of these curved parallels and meridians are of different sizes and shapes, complicating the location of points and the measurement of directions. To aid these essential operations, a rectangular grid is superimposed upon the projection. This grid (a series of straight lines intersecting at right angles) furnishes the map reader with a of squares similar to the block system of most city streets. The dimensions and orientation of different types of grids vary, but three

Amsterdam, Netherlands............4°53'01"E"
Athens, Greece.....................23°42'59"E
Batavia (Djakarta), Indonesia... 106°48'28"E
Bern, Switzerland...................7°26'22"E
Brussels, Belgium...................4°22'06"E
Copenhagen, Denmark............12°34'40"E
Ferro (Hierro), Canary Islands... 17°39'46"W
Helsinki, Finland..................24°53'17"E
Istanbul, Turkey...................28°58'50"E
Lisbon, Portugal....................9°07'55"W
Madrid, Spain......................3°41'15"W
Oslo, Norway......................10°43'23"E
Paris, France.........................2°20'14"E
Pulkovo, Russia....................30°19'39"E
Rome, Italy..........................12°27'08"E
Stockholm, Sweden...............18°03'30"E
Tirane, Albania.....................19°46'45"E

Table 1. Table of prime meridians.

properties are common to all military grid systems: one, they are true rectangular grids; two, they are superimposed on the geographic projection; and three, they permit linear and angular measurements.

a. **Universal Transverse Mercator Grid.** The UTM grid has been designed to cover that part of the world between latitude 84°N and latitude 80°S, and, as its name implies, is imposed on the transverse Mercator projection. Each of the 60 zones (6 degrees; wide) into which the globe is divided for the grid has its own origin at the intersection of its central meridian and the equator (Figure 4-8). The grid is identical in all 60 zones. Base

Figure 4-8. UTM grid zone location.

values (in meters) are assigned to the central meridian and the equator, and the grid lines are drawn at regular intervals parallel to these two base lines. With each grid line assigned a value denoting its distance from the origin, the problem of locating any point becomes progressively easier. Normally, it would seem logical to assign a value of zero to the two base lines and measure outward from them. This, however, would require either that directions—N, S, E, or W—be always given with distances, or that all points south of the equator or west of the central meridian have negative values. This inconvenience is eliminated by assigning "false values" to the base lines, resulting in positive values for all points within each zone. Distances are always measured RIGHT and UP (east and north as the reader faces the map), and the assigned values are called "false easting" and "false northing." (Figure 4-9) The false easting value for each central meridian is 500,000 meters, and the false north-ing value for the equator is O meters when measuring in the northern hemisphere and 10,000,000 meters when measuring in the southern hemisphere. The use of the UTM grid for point designation will be discussed in detail in paragraph 4-4.

b. **Universal Polar Stereographic Grid.** The UPS grid is used to represent the polar regions. (Figure 4-10)

(1) *North polar area* The origin of the UPS grid applied to the north polar area is

the north pole. The "north-south" base line is the line formed by the 0-degree and 180-degree meridians; the "east-west" base line is formed by the two 90-degree meridians.

(2) *South polar area.*The origin of the UPS grid in the south polar area is the south pole. The base lines are similar to those of the north polar area.

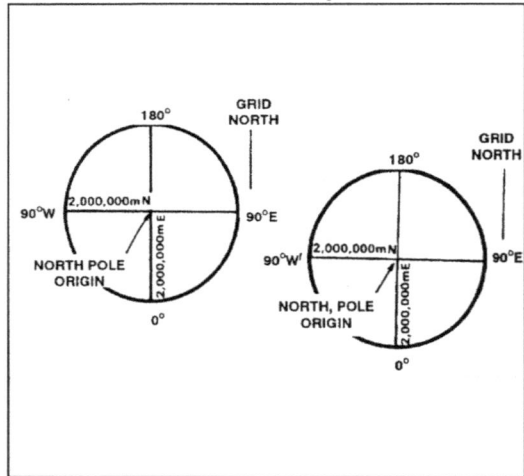

Figure 4-9. False eastings and northings for the UPS grid.

4-4. THE US ARMY MILITARY GRID REFERENCE: SYSTEM

This grid reference system is designated for use with the UTM and UPS grids. The coordinate value of points in these grids could contain as many as 15 digits if numerals alone were used. The US military grid reference system reduces the length of written coordinates by substituting single letters for several numbers. Using the UTM and the UPS grids, it is possible for the location of a point (identified by numbers alone) to be in many different places on the surface of the earth. With the use of the military grid reference system, there is no possibility of this happening.

a. **Grid Zone Designation.** The world is divided into 60 grid zones, which are large, regularly shaped geographic areas, each of which is given a unique identification called the grid zone designation.

(1) *UTM grid.* The first major breakdown is the division of each zone into areas 6° wide by 8° high and 6° wide by 12° high. Remember, for the transverse Mercator projection, the earth's surface between 80°S and 84°N is divided into 60 N-S zones, each 6° wide. These zones are numbered from west to east, 1 through 60, starting at the 180° meridian. This surface is divided into 20 east-west rows in which 19 are S° high and 1 row at the extreme north is 12° high. These rows are then lettered, from south to north, C through X (I and O were omitted). Any 6° by 8° zone or 6° by 12° zone can be identified by giving the number and

letter of the grid zone and row in which it lies. These are read RIGHT and UP so the number is always written before the letter. This combination of zone number and row letter constitutes the grid zone designation. Columbus lies in zone 16 and row S, or in grid zone designation 16S (Figure 4-8).

(2) *UPS grid.* The remaining letters of the alphabet, A, B, Y, and Z, are used for the UPS grids. Each polar area is divided into two zones separated by the 0-180° meridian. In the south polar area, the letter A is the grid zone designation for the area west of the 0-180° merid-ian, and B for the area to the east. In the north polar area, Y is the grid zone designation for the western area and Z for the eastern area (Figure 4-10).

b. **100,000-Meter Square.** Between 84°N and 80°S, each 6° by 8° or 6° by 12° zone is covered by 100,000-meter squares that are identified by the combination of two alphabetical letters. This identification is unique within the area covered by the grid zone designation. The first letter is the column designation; the second letter is the row designation (Page 4-10, Figure 4-11). The north and south polar areas are also divided into 100,000-meter squares by columns and rows. A detailed discussion of the polar system can be found in TM 8358.1. The 100,000-meter square identification letters are located in the grid reference box in the lower margin of the map.

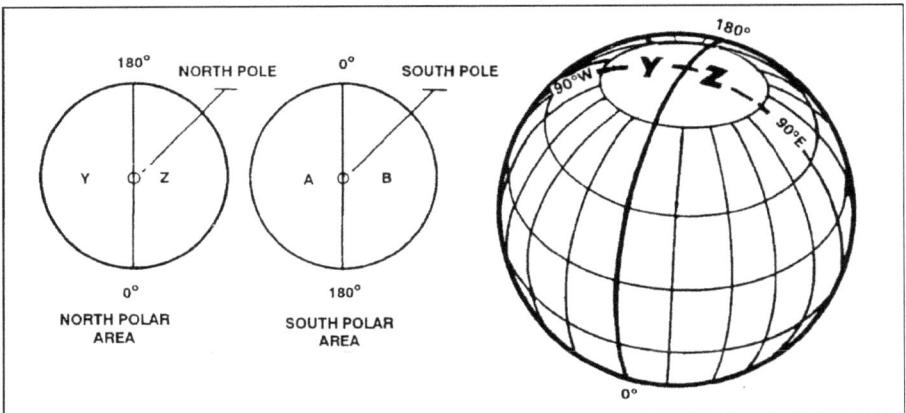

Figure 4-10. Grid zone designation for UPS grid.

PLATE 12

	96°			580,000m			90°			500,000m			84°	
QV	TQ	UQ	VQ	WQ	XQ	YQ	BV	CV	DV	EV	FV	GV	KQ		
QU	TP	UP	VP	WP	XP	YP	BU	CU	DU	EU	FU	GU	KP		
QT	TN	UN	VN	WN	XN	YN	BT	CT	DT	ET	FT	GT	KN		
QS	TM	UM	VM	WM	XM	YM	BS	CS	DS	ES	FS	GS	KM		
QR	TL	UL	VL	WL	XL	YL	BR	CR	DR	ER	FR	GR	K		
QQ	TK	UK	VK	WK	XK	YK	BQ	CQ	DQ	EQ	FQ	GQ	K		
QP	TJ	UJ	VJ	WJ	XJ	YJ	BP	CP	DP	EP	FP	GP	K		
QN	TH	UH	VH	WH	XH	YH	BN	CN	DN	EN	FN	GN	K		
QM	TG	UG	VG	WG	XG	YG	BM	CM	DM	EM	FM	GM	K		
QL	TF	UF	VF	WF	XF	YF	BL	CL	DL	EL	FL	GL	K		

Figure 4-11. Grid zone designation and 100,000-meter
square identification.

c. **Grid Coordinates.** We have now divided the earth's surface into 6° by 8° quadrangles, and covered these with 100,000-meter squares. The military grid reference of a point consists of the numbers and letters indicating in which of these areas the point lies, plus the coordinates locating the point to the desired position within the 100,000-meter square. The next step is to tie in the coordinates of the point with the larger areas. To do this, you must understand the following.

(1) *Grid lines*. The regularly spaced lines that make the UTM and the UPS grid on any large-scale maps are divisions of the 100,000-meter square; the lines are spaced at 10,000- or 1,000- meter intervals (Figure 4-12). Each of these lines is labeled at both ends of the map with its false easting or false northing value, showing its relation to the origin of the zone. Two digits of the values are printed in large type, and these same two digits appear at intervals along the grid lines on the face of the map. These are called the principal digits, and represent the 10,000 and 1,000 digits of the grid value. They are of major

importance to the map reader because they are the numbers he will use most often for referencing points. The smaller digits complete the UTM grid designation.

EXAMPLE: The first grid line north of the south west corner of the Columbus map is labeled 3570000m N. This means its false northing (distance north of the equator) is 3,570,000 meters. The principal digits, 70, identify the line for referencing points in the northerly direction. The smaller digits, 35, are part of the false coordinates and are rarely used. The last three digits, 000, of the value are omitted Therefore, the first grid line east of the south-west corner is labeled 689000m E. The principal digits, 89, identify the line for referencing points in the easterly direction (Figure 4-13).

(2) *Grid squares*. *The* north-south and east-west grid lines intersect at 90°, forming grid squares. Normally, the size of one of these grid squares on large-scale maps is 1,000 meters (1 kilometer).

Figure 4-12. Grid lines.

Figure 4-13. Columbus map, southwest corner.

(3) *Grid coordinate scales.* The primary tool for plotting grid coordinates is the grid coordinate scale. The grid coordinate scale divides the grid square more accurately than can be done by estimation, and the results are more consistent. When used correctly, it presents less chance for making errors. GTA 5-2-12, 1981, contains four types of coordinate scales (Figure 4-14).

(a) The 1:25,000/1:250,000 (lower right in figure) can be used in two different scale maps, 1:25,000 or 1:250,000. The 1:25,000 scale subdivides the 1,000-meter grid block into 10 major subdivisions, each equal to 100 meters. Each 100-meter block has five graduations, each equal to 20 meters. Points falling between the two graduations can be read accurately by the use of estimation. These values are the fourth and eighth digits of the coordinates. Likewise, the 1:250,000 scale is subdivided in 10 major subdivisions, each equal to 1,000 meters. Each 1,000-meter block has five graduations, each equal to 200 meters. Points falling between two graduations can be read approximately by the use of estimation.

(b) The 1:50,000 scale (upper left in figure) subdivides the 1,000-meter block into 10 major subdivisions,

each equal to 100 meters. Each 100 meter block is then divided in half. Points falling between the graduations must be estimated to the nearest 10 meters for the fourth and eighth digits of the coordinates.

(c) The 1:100,000 scale (lower left in figure) subdivides the 1,000-meter grid block into five major subdivisions of 200 meters each. Each 200-meter block is then divided in half at 100-meter intervals.

4-5. LOCATING A POINT USING GRID COORDINATES

Based on the military principle for reading maps (RIGHT and UP), locations on the map can be determined by grid coordinates. The number of digits represents the degree of precision to which a point has been located and measured on a map—the more digits the more precise the measurement.

a. **Without a Coordinate Scale.** In order to determine grids without a coordinate scale, the reader simply refers to the north-south grid lines numbered at the bottom margin of any map. Then he reads RIGHT to the north-south grid line that precedes the desired point (this first set of two digits is the RIGHT reading). Then by referring to the east-west grid lines numbere d at either side of the map, the map reader moves UP to the east-west grid line that precedes the desired point (these two digits are the UP reading). Coordinates 1484 locate the 1,000-meter grid square in which point X is located, the next square to the right would be 1584; the next square up would be 1485, and so forth (Figure 4-15). To locate the point to the nearest 100 meters, use estimation. By mentally dividing the grid square in tenths, estimate the distance from the grid line to the point in the same order (RIGHT and UP). Give complete coordinate RIGHT, then complete coordinate UP. Point X is about two-tenths or 200 meters to the RIGHT into the grid square and about seven-tenths or 700 meters UP. The coordinates to the nearest 100 meters are 142847.

Figure 4-14. Coordinate scales.

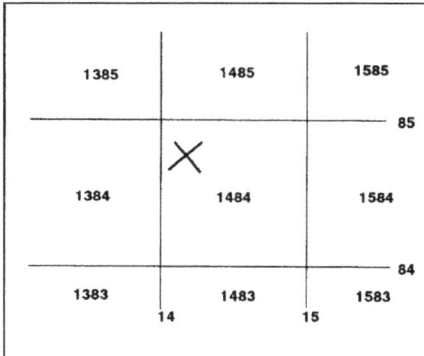

Figure 4-15. Determining grids without
coordinate point.

b. **With a Coordinate Scale.** In order to use the coordinate scale for determining grid coordinates, the map user has to make sure that the appropriate scale is being used on the corresponding map, and that the scale is right side up. To ensure the scale is correctly aligned, place it with the zero-zero point at the lower left corner of the grid square. Keeping the horizontal line of the scale directly on top of the east-west grid line, slide it to the right until the vertical line of the scale touches the point for which the coordinates are desired (Figure 4-16). When reading coordinates, examine the two sides of the coordinate scale to ensure that the horizontal line of the scale is aligned with the east-west grid line, and the vertical line of the scale is parallel with the north-south grid line. The scale is used when precision of more than 100 meters is required. To locate the point to the nearest 10 meters, measure the hundredths of a grid square RIGHT and UP from the grid lines to the point. Point X is about 17 hundredths or 170 meters RIGHT and 84 hundredths or 840 meters UP. The coordinates to the nearest 10 meters are 14178484.

c. **Recording and Reporting Grid Coordinates.** Coordinates are written as one continuous number without spaces, parentheses, dashes, or decimal points; they must always contain an even number of digits. Therefore, whoever is to use the written coordinates must know where to make the split between the RIGHT and UP readings. It is a military requirement that the 100,000-meter square identification letters be included in any point designation. Normally, grid coordinates are determined to the nearest 100 meters (six digits) for reporting locations. With practice, this can be done without using plotting scales. The location of targets and other point locations for fire support are determined to the nearest 10 meters (eight digits).

Figure 4-16. Placing a coordinate scale on a grid.

NOTE: Refer to Figure 4-17. Care should be exercised by the map reader using the coordinate scale when the desired point is located within the zero-zero point and the number 1 on the scale. Always prefix a zero if the hundredths reading is less than 10. In Figure 4-17, the desired point should be reported as 14818407.

Figure 4-17. Zero-zero point.

NOTE: Special care should be exercised when recording and reporting coordinates. Transposing numbers or making errors could be detrimental to military operations.

4-6. LOCATING A POINT USING THE US ARMY MILITARY GRID REFERENCE SYSTEM

There is only one rule to remember when reading or reporting grid coordinates—always read to the RIGHT and then UP. The first half of the reported set of coordinate digits represents the left-to-right (easting) grid label, and the second half represents the label as read from the bottom to top (northing). The grid coordinates may represent the location to the nearest 10-, 100-, or 1,000-meter increment.

a. **Grid Zone.** The number 16 locates a point within zone 16, which is an area 6° wide and extends between 80°S latitude and 84°N latitude (Figure 4-8).

b. **Grid Zone Designation.** The number and letter combination, 16S, further locates a point

within the grid zone designation 16S,which is a quadrangle 6° wide by 8° high. There are 19 of these quads in zone 16. Quad X, which is located between 72°N and 84°N latitude, is 12° high (Figure 4-8).

c. **100,000-Meter Square Identification.** The addition of two more letters locates a point within the 100,000-meter grid square. Thus 16SGL (Figure 4-11) locates the point within the 100,000-meter square GL in the grid zone designation 16S. For information on the lettering system of 100,000-meter squares, see TM 5-241-1.

d. **10,000-Meter Square.** The breakdown of the US Army military grid reference system continues as each side of the 100,000-meter square is divided into 10 equal parts. This division produces lines that are 10,000 meters apart. Thus the coordinates 16SGL08 would locate a point as shown in Figure 4-18A The 10,000-meter grid lines appear as index (heavier) grid lines on maps at 1:100,000 and larger.

Figure 4-18A. The 10,000-meter grid square

Figure 4-18B. The 1,000-meter grid square.

e. **1,000-Meter Square.** To obtain 1,000-meter squares, each side of the 10,000-meter square is divided into 10 equal parts. This division appears on large-scale maps as the actual grid lines, they are 1,000 meters apart. On the Columbus map, using coordinates 16SGL0182, the easting 01 and the northing 82 gives the location of the south-west corner of grid square 0182 or to the nearest 1,000 meters of a point on the map (Figure 4-18B).

f. **100-Meter Identification.** To locate to the nearest 100 meters, the grid coordinate scale can be used to divide the 1,000-meter grid squares into 10 equal parts

g. **10-Meter Identification.** The grid coordinate scale has divisions every 50 meters on the 1:50,000 scale and every 20 meters on the 1:25,000 scale. These can be used to estimate to the nearest 10 meters and give the location of one point on the earth's surface to the nearest 10 meters.

Example:
16SGLO1948253
(gas tank) (Figure 4-18C).

h. **Precision.** The precision of a point's location is shown by the number of digits in the coordinates; the more digits, the more precise the location (Figure 4-18C, insert).

Figure 4-18-C. The 100-meter and 10-meter grid squares.

4-7. GRID REFERENCE BOX

A grid reference box (Figure 4-19) appears in the marginal information of each map sheet. It contains step-by-step instructions for using the grid and the US Army military grid reference system. The grid reference box is divided into two parts.

a. The left portion identifies the grid zone designation and the 100,000-meter square. If the sheet falls in more than one 100,000-meter square, the grid lines that separate the squares are shown in the diagram and the letters identifying the 100,000-meter squares are given.

EXAMPLE: On the Columbus map sheet, the vertical line labeled 00 is the grid line that separates the two 100,000-meter squares, FL and GL. The left portion also shows a sample for the 1,000-meter square with its respective labeled grid coordinate numbers and a sample point within the 1,000-meter square.

b. The right portion of the grid reference box explains how to use the grid and is keyed on the sample 1,000-meter square of the left side. The following is an example of the military grid reference:

EXAMPLE: 16S locates the 6° by 8° area (grid zone designation).

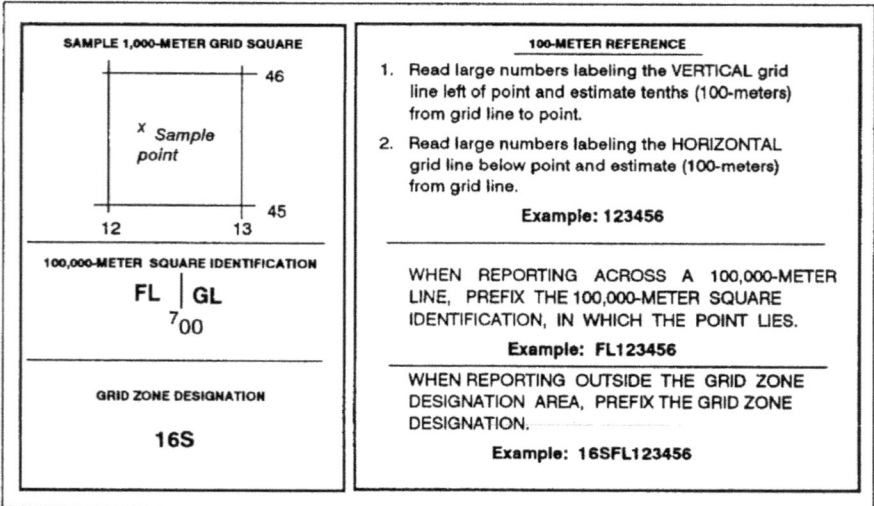

SAMPLE 1,000-METER GRID SQUARE	100-METER REFERENCE
46 *x Sample point* 45 12 13	1. Read large numbers labeling the VERTICAL grid line left of point and estimate tenths (100-meters) from grid line to point. 2. Read large numbers labeling the HORIZONTAL grid line below point and estimate (100-meters) from grid line. **Example: 123456**
100,000-METER SQUARE IDENTIFICATION FL $\mid$ GL ^{7}00	WHEN REPORTING ACROSS A 100,000-METER LINE, PREFIX THE 100,000-METER SQUARE IDENTIFICATION, IN WHICH THE POINT LIES. **Example: FL123456**
GRID ZONE DESIGNATION 16S	WHEN REPORTING OUTSIDE THE GRID ZONE DESIGNATION AREA, PREFIX THE GRID ZONE DESIGNATION. **Example: 16SFL123456**

Figure 4-19. Grid reference box.

4-8. OTHER GRID SYSTEMS

The military grid reference system is not universally used. You must be prepared to interpret and use other grid systems, depending on your area of operations or the personnel you are operating with.

a. **British Grids.** In a few areas of the world, British grids are still shown on military maps. However, the British grid systems are being phased out. Eventually all military mapping will be converted to the UTM grid.

b. **The World Geographic Reference System (GEOREF).** This is a worldwide position reference system used primarily by the US Air Force. It may be used with any map or chart that has latitude and longitude printed on it. Instructions for using GEOREF data are printed in blue and are found in the margin of aeronautical charts (Figure 4-20). This system is based upon a division of the earth's surface into quadrangles of latitude and longitude having a systematic identification code. It is a method of expressing latitude and longitude in a form suitable for rapid reporting and plotting. Figure 4-20 illustrates a sample grid reference box using GEOREF. The GEOREF system uses an identification code that has three main divisions.

(1) *First division.* There are 24 north-south (longitudinal) zones, each 15° wide. These zones, starting at 180° and progressing eastward, are lettered A through Z (omitting I and O). The first letter of any GEOREF coordinate identifies the north-south zone in which the point is located. There are 12 east-west (latitudinal) bands, each 15° wide. These bands are lettered A through M (omitting I) northward from the south pole. The second letter of any GEOREF coordinate identifies the east-west band in which the point is located. The zones and bands divide the earth's surface into 288 quadrangles, each identified by two letters.

(2) *Second division.* Each 15° quadrangle is further divided into 225 quadrangles of 1° each (15° by 15°). This division is effected by dividing a basic 15° quadrangle into 15 north-south zones and 15 east-west bands. The north-south zones are lettered A through Q (omitting I and O) from west to east. The third letter of any GEOREF coordinate identifies the 1° north-south zone within a 15° quadrangle. The east-west bands are lettered A through Q (I and O omitted)

from south to north. The fourth letter of a GEOREF coordinate identifies the 1° east-west band within a 15° quadrangle. Four letters will identify any 1° quadrangle in the world.

(3) *Third division.* Each of the 1° quadrangles is divided into 3,600 one-minute quadrangles. These one-minute quadrangles are formed by dividing the 1° quadrangles into 60 one-minute north-south zones numbered 0 through 59 from west to east, and 60 east-west bands numbered 0 to 59 from south to north. To designate any one of the 3,600 one-minute quadrangles requires four letters and four numbers. The rule READ RIGHT AND UP is always followed. Numbers 1 through 9 are written as 01, 02, and so forth. Each of the 1-minute quadrangles may be further divided into 10 smaller divisions both north-south and east-west, permitting the identification of 0.1-minute quadrangles. The GEOREF coordinate for any 0.1-minute quadrangle consists of four letters and six numbers.

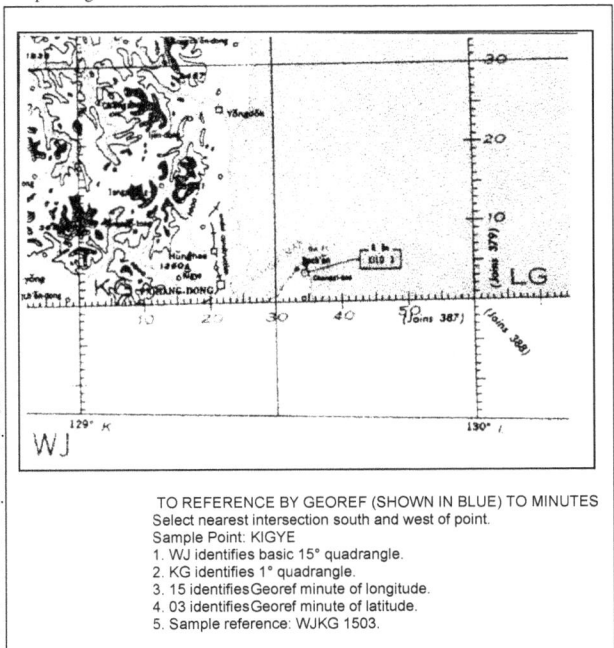

TO REFERENCE BY GEOREF (SHOWN IN BLUE) TO MINUTES
Select nearest intersection south and west of point.
Sample Point: KIGYE
1. WJ identifies basic 15° quadrangle.
2. KG identifies 1° quadrangle.
3. 15 identifies Georef minute of longitude.
4. 03 identifies Georef minute of latitude.
5. Sample reference: WJKG 1503.

Figure 4-20. Sample reference using GEOREF

4-9 PROTECTION OF MAP COORDINATES AND LOCATIONS

A disadvantage of any standard system of location is that the enemy, if he intercepts one of our messages using the system, can interpret the message and find our location. This possibility can be eliminated by using an authorized low-level numerical code to express locations. Army Regulation 380-40 outlines the procedures for obtaining authorized codes.

a. The authorized numerical code provides a capability for encrypting map references and other numerical information that requires short-term security protection when, for operational reasons, the remainder of the message is transmitted in plain language. The system is published in easy-to-use booklets with sufficient material in each for one month's operation. Sample training editions of this type of system are available through the unit's communications and electronics officer.

b. The use of any encryption methods other than authorized codes is, by regulation, unauthorized and shall not be used.

SCALE AND DISTANCE

A map is a scaled graphic representation of a portion of the earth's surface. The scale of the map permits the user to convert distance on the map to distance on the ground or vice versa. The ability to determine distance on a map, as well as on the earth's surface, is an important factor in planning and executing military missions.

5-1. REPRESENTATIVE FRACTION

The numerical scale of a map indicates the relationship of distance measured on a map and the corresponding distance on the ground. This scale is usually written as a fraction and is called the representative fraction. The RF is always written with the map distance as 1. It is independent of any unit of measure. (It could be yards, meters, inches, and so forth.) An RF of 1/50,000 or 1:50,000 means that one unit of measure on the map is equal to 50,000 units of the same measure on the ground.

a. The ground distance between two points is determined by measuring between the same two points on the map and then multiplying the map measurement by the denominator of the RF or scale (Figure 5-1).

EXAMPLE:

The map scale is 1:50,000

RF = 1/50,000

The map distance from point A to point B is 5 units

5 x 50,000 = 250,000 units of ground distance

b. Since the distance on most maps is marked in meters and the RF is expressed in this unit of measurement in most cases, a brief description of the metric system is needed. In the metric system, the standard unit of measurement is the meter.

1 meter contains 100 centimeters (cm).

100 meters is a regular football field plus 10 meters.

1,000 meters is 1 kilometer (km).

10 kilometers is 10,000 meters.

Appendix C contains the conversion tables.

c. The situation may arise when a map or sketch has no RF or scale. To be able to determine ground distance on such a map, the RF must be determined. There are two ways to do this:

(1) *Comparison with ground distance.*

(a) Measure the distance between two points on the map—map distance (MD).

(b) Determine the horizontal distance between these same two points on the ground—ground distance (GD).

(c) Use the RF formula and remember that RF must be in the general form:

$$RF = \frac{1}{X} = \frac{MD}{GD}$$

(d) Both the MD and the GD must be in the same unit of measure and the MD must be reduced to 1.

EXAMPLE:

MD = 4.32 centimeters

GD = 2.16 kilometers
(216,000 centimeters)

$$RF = \frac{1}{X} = \frac{4.32}{216,000}$$

or

$$\frac{216,000}{4.32} = 50,000$$

therefore

$$RF = \frac{1}{50,000} \quad or \ 1:50,000$$

Figure 5-1. Converting map distance to ground distance.

(2) *Comparison with another map of the same area that has an RF.*

(a) Select two points on the map with the unknown RF. Measure the distance (MD) between them.

(b) Locate those same two points on the map that have the known RF. Measure the distance (MD) between them. Using the RF for this map, determine GD, which is the same for both maps.

(c) Using the GD and the MD from the first map, determine the RF using the formula:

$$RF = \frac{1}{X} = \frac{MD}{GD}$$

d. Occasionally it may be necessary to determine map distance from a known ground distance and the RF:

$$MD = \frac{GD}{Denominator\ or\ RF}$$

Ground Distance = 2, 200 meters

RF = 1:50,000

$$MD = \frac{2,200\ meters}{50,000}$$

MD = 0.044 meter x 100(centimeters per meter)

MD = 4.4 centimeters

e. When determining ground distance from a map, the scale of the map affects the accuracy. As the scale becomes smaller, the accuracy of measurement decreases because some of the features on the map must be exaggerated so that they may be readily identified.

5-2. GRAPHIC (BAR) SCALES

A graphic scale is a ruler printed on the map and is used to convert distances on the map to actual ground distances. The graphic scale is divided into two parts. To the right of the zero, the scale is marked in full units of measure and is called the primary scale. To the left of the zero, the scale is divided into tenths and is called the extension scale. Most maps have three or more graphic scales, each using a different unit of measure (Figure 5-2). When using the graphic scale, be sure to use the correct scale for the unit of measure desired.

a. To determine straight-line distance between two points on a map, lay a straight-edged piece of paper on the map so that the edge of the paper touches both points and extends past them. Make a tick mark on the edge of the paper at each point (Figure 5-3).

b. To convert the map distance to ground distance, move the paper down to the graphic bar scale, and align the right tick mark (b) with a printed number in the primary scale so that the left tick mark (a) is in the extension scale (Figure 5-4).

c. The right tick mark (b) is aligned with the 3,000-meter mark in the primary scale, thus the distance is at least 3,000 meters. To determine the distance between the two points to the nearest 10 meters, look at the extension scale. The extension scale is numbered with zero at the right and increases to the left. When using the extension scale, always read right to left (Figure 5-4). From the zero left to the end of the first shaded square is 100 meters. From the beginning of the center square to the left is 100 to 200 meters; at the beginning of the second shaded square is 200 to 300 meters. Remember, the distance in the extension scale increases from right to left.

d. To determine the distance from the zero to tick mark (a), divide the distance inside the squares into

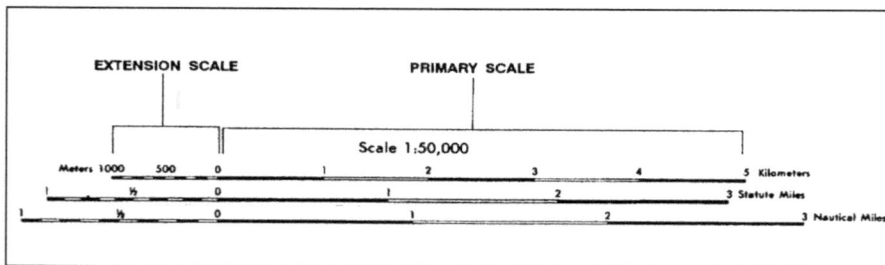

Figure 5-2. Using a graphic (bar) scale.

NOTE: When measuring distance, use center mass of symbol or object.

Figure 5-3. Transferring map distance to paper strip.

tenths (Figure 5-4). As you break down the distance between the squares in the extension scale into tenths, you will see that tick mark (a) is aligned with the 950-meter mark. Adding the distance of 3,000 meters determined in the primary scale to the 950 meters you determined by using the extension scale, we find that the total distance between points (a) and (b) is 3,950 meters.

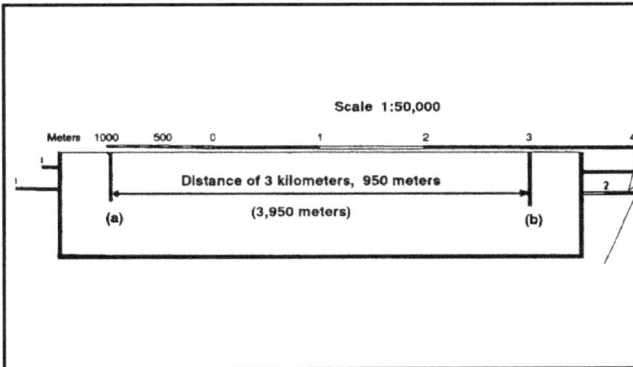

Figure 5-4. Measuring straight-line map distance.

e. To measure distance along a winding road, stream, or other curved line, the straight edge of a piece of paper is used. In order to avoid confusion concerning the point to begin measuring from and the ending point, an eight-digit coordinate should be given for both the starting and ending points. Place a tick mark on the paper and map at the beginning point from which the curved line is to be measured. Align the edge of the paper along a straight portion and make a tick mark on both map and paper when the edge of the paper leaves the straight portion of the line being measured (Figure 5-5A).

f. Keeping both tick marks together (on paper and map), place the point of the pencil close to the edge of the paper on the tick mark to hold it in place and pivot the paper until another straight portion of the curved line is aligned with the edge of the paper. Continue in this manner until the measurement is completed (Figure 5-5B).

g. When you have completed measuring the distance, move the paper to the graphic scale to determine the ground distance. The only tick marks you will be measuring the distance between are tick marks (a) and (b). The tick marks in between are not used (Figure 5-5C).

Figure 5-5. Measuring a curved line.

h. There may be times when the distance you measure on the edge of the paper exceeds the graphic scale.
In this case, there are different techniques you can use to
determine the distance.

(1) One technique is to align the right tick mark (b)
with a printed number in the primary scale, in this case
the 5. You can see that from point (a) to point (b) is
more than 6,000 meters when you add the 1,000 meters
in the extension scale. To determine the exact distance
to the nearest 10 meters, place a tick mark (c) on the
edge of the paper at the end of the extension scale
(Figure 5-6A). You know that from point (b) to point (c)
is 6,000 meters. With the tick mark (c) placed on the
edge of the paper at the end of the extension scale,
slide the paper to the right. Remember the distance in the
extension is always read from right to left. Align tick
mark (c) with zero and then measure the distance
between tick marks (a) and (c). The distance between
tick marks (a) and (c) is 420 meters. The total ground
distance between start and finish points is 6,420 meters
(Figure 5-6B).

Figure 5-6. Determining the exact distance.

(2) Another technique that may be used to determine exact distance between two points when the edge of the paper exceeds the bar scale is to slide the edge of the paper to the night until tick mark (a) is aligned with the edge of the extension scale. Make a tick mark on the paper, in line with the 2,000 meter mark (c) (Figure 5-7A). Then slide the edge of the paper to the left until tick mark (b) is aligned with the zero. Estimate the 100-meter increments into 10-meter increments to determine how many meters tick mark (c) is from the zero line (Figure 5-7B). The total distance would be 3,030 meters.

(3) At times you may want to know the distance from a point on the map to a point off the map. In order to do this, measure the distance from the start point to the edge of the map. The marginal notes give the road distance from the edge of the map to some towns, highways, or junctions off the map. To determine the total distance, add the distance measured on the map to the distance given in the marginal notes. Be sure the unit of measure is the same.

(4) When measuring distance in statute or nautical miles, round it off to the nearest one-tenth of a mile and make sure the appropriate bar scale is used.

(5) Distance measured on a map does not take into consideration the rise and fall of the land. All distances measured by using the map and graphic scales are flat distances. Therefore, the distance measured on a map will increase when actually measured on the ground. This must be taken into consideration when navigating across country.

i. The amount of time required to travel a certain distance on the ground is an important factor in most military operations. This can be determined if a map of the area is available and a graphic time-distance scale is constructed for use with the map as follows:

$R = Rate\ of\ travel\ (speed)$ $\qquad$ $T = Time$

$D = Distance\ (ground\ distance)$ $\qquad$ $T = \dfrac{D}{R}$

Figure 5-7. Reading the extension scale.

For example, if an infantry unit is marching at an average rate (R) of 4 kilometers per hour, it will take approximately 3 hours (T) to travel 12 kilometers.

$$\frac{12(D)}{4(R)} = 3(T)$$

j. To construct a time-distance scale (Figure 5-8A), knowing your length of march, rate of speed, and map scale, that is, 12 kilometers at 3 kilometers per hour on a 1:50,000-scale map, use the following process:

(1)Mark off the total distance on a line by referring to the graphic scale of the map or, if this is impracticable, compute the length of the line as follows:

(a) Convert the ground distance to centimeters: 12 kilometers x 100,000 (centimeters per kilometer) = 1,200,000 centimeters.

(b) Find the length of the line to represent the distance at map scale —

$$MD = \frac{1}{50,000} = \frac{1,200,000\,cm}{50,000} = 24\,centimeters$$

(c) Construct a line 24 centimeters in length. (Figure5-8A)

(2) Divide the line by the rate of march into three parts (Figure 5-8B), each part representing the distance traveled in one hour, and label.

(3) Divide the scale extension (left portion) into the desired number of lesser time divisions—

1-minute divisions — 60

5-minute divisions — 12

10-minute divisions— 6

(4) Figure 5-8C shows a 5-minute interval scale. Make these divisions in the same manner as for a graphic scale. The completed scale makes it possible to determine where the unit will be at any given time. However, it must be remembered that this scale is for one specific rate of march only, 4 kilometers per hour.

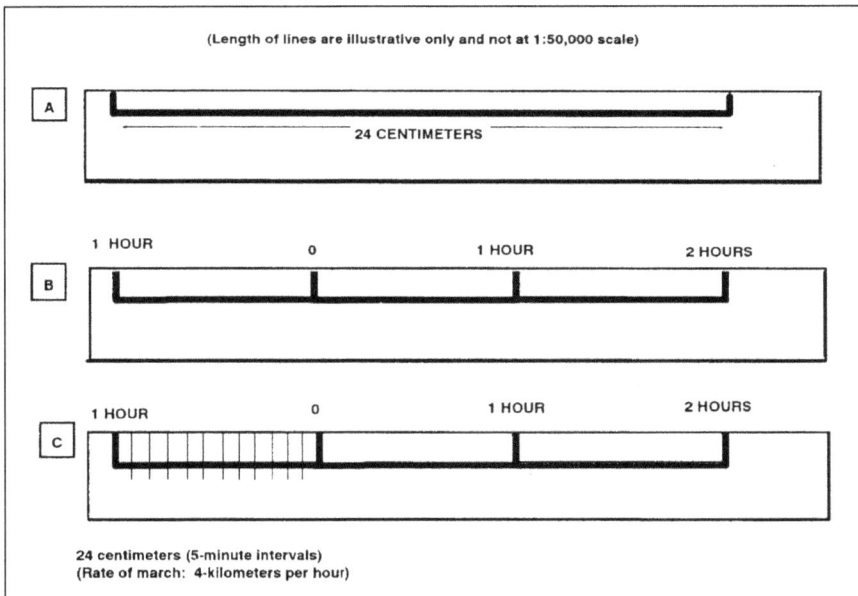

Figure 5-8. Constructing a time-distance scale.

5-3. OTHER METHODS

Determining distance is the most common source of error encountered while moving either mounted or dismounted. There may be circumstances where you are unable to determine distance using your map or where you are without a map. It is therefore essential to learn methods by which you can accurately pace, measure, use subtense, or estimate distances on the ground.

a. **Pace Count.** Another way to measure ground distance is the pace count. A pace is equal to one natural step, about 30 inches long. To accurately use the pace count method, you must know how many paces it takes you to walk 100 meters. To determine this, you must walk an accurately measured course and count the number of paces you take. A pace course can be as short as 100 meters or as long as 600 meters. The pace course, regardless of length, must be on similar terrain to that you will be walking over. It does no good to walk a course on flat terrain and then try to use that pace count on hilly terrain. To determine your pace count on a 600-meter course, count the paces it takes you to walk the 600 meters, then divide the total paces by 6. The answer will give you the average paces it takes you to walk 100 meters. It is important that each person who navigates while dismounted knows his pace count.

(1) There are many methods to keep track of the distance traveled when using the pace count. Some of these methods are: put a pebble in your pocket every time you have walked 100 meters according to your pace count; tie knots in a string; or put marks in a notebook. Do not try to remember the count; always use one of these methods or design your own method.

(2) Certain conditions affect your pace count in the field, and you must allow for them by making adjustments.

(a) *Slopes*. Your pace will lengthen on a downslope and shorten on an upgrade. Keeping this in mind, if it normally takes you 120 paces to walk 100 meters, your pace count may increase to 130 or more when walking up a slope.

(b) *Winds*. A head wind shortens the pace and a tail wind increases it.

(c) *Surfaces*. Sand, gravel, mud, snow, and similar surface materials tend to shorten the pace.

(d) *Elements*. Falling snow, rain, or ice cause the pace to be reduced in length.

(e) *Clothing*. Excess clothing and boots with poor traction affect the pace length.

(f) *Visibility*. Poor visibility, such as in fog, rain, or darkness, will shorten your pace.

b. **Odometer.** Distances can be measured by an odometer, which is standard equipment on most vehicles. Readings are recorded at the start and end of a course and the difference is the length of the course.

(1) To convert kilometers to miles, multiply the number of kilometers by 0.62.

EXAMPLE:

16 *kilometers* **= 16 x 0.62 = 9.92** *miles*

(2) To convert miles to kilometers, devide the number of miles by 0.62.

EXAMPLE:

10 miles = 10 *divided by* **0.62 = 16.77 kilometers**

c. **Subtense.** The subtense method is a fast method of determining distance and yields accuracy equivalent to that obtained by measuring distance with a premeasured piece of wire. An advantage is that a horizontal distance is obtained indirectly; that is, the distance is computed rather than measured. This allows subtense to be used over terrain where obstacles such as streams, ravines, or steep slopes may prohibit other methods of determining distance.

(1) The principle used in determining distance by the subtense method is similar to that used in estimating distance by the mil relation formula. The field artillery application of the mil relation formula involves only estimations. It is not accurate enough for survey purposes. However, the subtense method uses precise values with a trigonometric solution. Subtense is based on a principle of visual perspective—the farther away an object, the smaller it appears.

(2) The following two procedures are involved in subtense measurement:

•Establishing a base of known length.

•Measuring the angle of that base by use of the aiming circle.

(3) The subtense base may be any desired length. However, if a 60-meter base, a 2-meter bar, or the length of an M16A1 or M16A2 rifle is used, precomputed subtense tables are available. The M16 or 2-meter bar must be held horizontal and perpendicular to the line of sight by a soldier facing the aiming circle. The instrument operator sights on one end of the M16 or 2-meter bar and measures the horizontal clockwise angle to the other end of the rifle or bar. He does this twice and averages the angles. He then enters the appropriate subtense table with the mean angle and extracts the distance. Accurate distances can be obtained with the M16 out to approximately 150 meters, with the 2-meter bar out to 250 meters, and with the 60-meter base out to 1,000 meters. If a base of another length is desired, a

distance can be computed by using the following formula:

$$Distance = \frac{\frac{1}{2} \ (base \ in \ meters)}{\tan \ (\frac{1}{2}) \ (in \ mils)}$$

d. **Estimation.** At times, because of the tactical situation, it may be necessary to estimate range. There are two methods that may be used to estimate range or distance.

(1) *100-meter unit-of-measure method*. To use this method, the soldier must be able to visualize a distance of 100 meters on the ground. For ranges up to 500 meters, he determines the number of 100-meter increments between the two objects he wishes to measure. Beyond 500 meters, the soldier must select a point halfway to the object(s) and determine the number of 100-meter increments to the halfway point, then double it to find the range to the object(s) (Figure 5-9).

(2) *Flash-to-bang method*. To use this method .to determine range to an explosion or enemy fire, begin to count when you see the flash. Count the seconds until you hear the weapon fire. This time interval may be measured with a stopwatch or by using a steady count, such as one-thousand-one, one-thousand-two, and so forth, for a three-second estimated count. If you must count higher than 10 seconds, start over with one. Multiply the number of seconds by 330 meters to get the approximate range (FA uses 350 meters instead).

Figure 5-9. Using a 100-meter unit-of-measure method.

(3) *Proficiency of methods*. The methods discussed above are used only to estimate range (Table 5-1). Proficiency in both methods requires constant practice. The best training technique is to require the soldier to pace the range after he has estimated the distance. In this way, the soldier discovers the actual range for himself, which makes a greater impression than if he is simply told the correct range.

Factors Affecting Range Estimation	Factors Causing Underestimation Range	Factors Causing Overestimation of Range
The clearness of outline and details of the object.	When most of the object is visible and offers a clear outline. surroundings.	When only a small part of the object can be seen or the object is small in relation to its
Nature of terrain or position of the observer.	When looking across a depression that is mostly hidden from view' When lookingdownward from high ground. When looking down a straight, open road or along a railroad. When looking over uniform surfaces like water, snow, desert, or grain fields. In bright light or when the sun is shining from behind the observer.	When looking across a depression that is totally visible. When vision is confined, as in streets, draws, or forrest trails. When looking from low ground toward high ground. In poor light, such as dawn and dusk; in rain, snow, fog; or when the sun is in the observer's eyes
Light and atmosphere.	When the object is insharp contest with the background or is silhouetted because of its size, shape, or color. When seen in the clear air of high altitudes.	When object blends into the background or terrain.

Table 5-1. Factors of range estimation.

DIRECTION

Being in the right place at the prescribed time is necessary to successfully accomplish military missions. Direction plays an important role in a soldier's everyday life. It can be expressed as right, left, straight ahead, and so forth; but then the question arises, "To the right of what?" This chapter contains the definition of azimuth and the three different norths, how to determine grid and magnetic azimuths with the use of the protractor and the compass, the use of some field-expedient methods to find directions, the declination diagram, and the conversion of azimuths from grid to magnetic and vice versa. It also includes some advanced aspects of map reading, such as intersection, resection, modified resection,, and polar plots.

6-1. METHODS OF EXPRESSING DIRECTION

Military personnel need a way of expressing direction that is accurate, is adaptable to any part of the world, and has a common unit of measure. Directions are expressed as units of angular measure.

a. **Degree.** The most common unit of measure is the degree (°) with its subdivisions of minutes (') and seconds (").

1 degree = 60 minutes.
1 minute = 60 seconds.

b. **Mil.** Another unit of measure, the mil (abbreviated ⫫m), is used mainly in artillery, tank, and mortar gunnery. The mil expresses the size of an angle formed when a circle is divided into 6,400 angles with the vertex of the angles at the center of the circle. A relationship can be established between degrees and mils. A circle equals 6400 mils divided by 360 degrees, or 17.78 mils per degree. To convert degrees to mils, multiply degrees by 17.78.

c. **Grad.** The grad is a metric unit of measure found on some foreign maps. There are 400 grads in a circle (a 90° right angle equals 100 grads). The grad is divided into 100 centesimal minutes (centigrade) and the minute into 100 centesimal seconds (milligrads).

6-2. BASE LINES

In order to measure something, there must always be a starting point or zero measurement. To express direction as a unit of angular measure, there must be a starting point or zero measure and a point of reference. These two points designate the base or reference line. There are three base lines—true north, magnetic north, and grid north. The most commonly used are magnetic and grid.

a. **True North.** A line from any point on the earth's surface to the north pole. All lines of longitude are true north lines. True north is usually represented by a star (Figure 6-1).

b. **Magnetic North.** The direction to the north magnetic pole, as indicated by the north-seeking needle of a magnetic instrument. Magnetic north is usually symbolized by a line ending with a half arrowhead (Figure 6-1). Magnetic readings are obtained with magnetic instruments, such as lensatic and M2 compasses.

c. **Grid North.** The north that is established by using the vertical grid lines on the map. Grid north may be symbolized by the letters GN or the letter ϒ" (Figure 6-1).

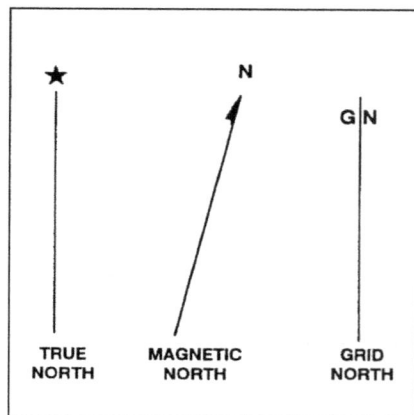

Figure 6-1. Three norths.

6. AZIMUTHS

 azimuth is defined as a horizontal angle measured
ckwise from a north base line. This north base line
ıld be true north, magnetic north, or grid north. The
muth is the most common military method to
press direction. When using an azimuth, the point
m which the azimuth originates is the center of an
aginary circle (Figure 6-2). This circle is divided
o 360° or 6400 mils (see Appendix G).

Figure 6-2. Origin of azimuth circle.

a. **Back Azimuth.** A back azimuth is the opposite
ection of an azimuth. It is comparable to doing an
oout face." To obtain a back azimuth from an
muth add 180° if the azimuth is 180° or less; or
otract 180° if the azimuth is 180° or more (Figure
6). The back azimuth of 180° may be stated as 0° or
mils, if th azimuth is less than 3200 mils, add 3200
Is; if the azimuth is more than 3200 mils, subtract
00 mils.

> **WARNING**
> When converting azimuths into back azi-
> muths, extreme care should be exercised
> when adding or subtracting the 180°. A
> simple mathematical mistake could
> cause disastrous consequences.

b. **Magnetic Azimuth.** The magnetic azimuth is
determined by using magnetic instruments, such as
lensatic and M-2 compasses. Refer to Chapter 9, paragraph 4, for details.

c. **Field-Expedient Methods.** Several field-expedient
methods to determine direction are discussed in Chapter 9,
paragraph 5.

6-4. GRID AZIMUTHS
When an azimuth is plotted on a map between point A
(starting point) and point B (ending point), the points are
joined together by a straight line. A protractor is used to
measure the angle between grid north and the drawn line,
and this measured azimuth is the grid azimuth (Figure
6-4).

> **WARNING**
> When measuring azimuths on a map, remem-
> ber that you are measuring from a starting
> point to an ending point. If a mistake is made
> and the reading is taken from the ending
> point, the grid azimuth will be opposite, thus
> causing the user to go in the wrong direction.

EXAMPLE:				
Azimuth	±	Back Azimuth		
115°	+180°	=	295°	
337°=	-180°	=	157°	

Figure 6-3. Back azimuth.

Figure 6-4. Measuring an azimuth.

5. PROTRACTOR

here are several types of protractors—full circle, half circle, square, and rectangular (Figure 6-5). All of them divide the circle into units of angular measure, and each has a scale around the outer edge and an index mark. The index mark is the center of the protractor circle from which all directions are measured.

a. The military protractor, GTA 5-2-12, contains two scales; one in degrees (inner scale) and one in mils (outer scale). This protractor represents the azimuth circle. The degree scale is graduated from 0° to 360°; each tick mark on the degree scale represents one degree. A line from 0° to 180° is called the base line of the protractor. Where the base line intersects the horizontal line, between 90° and 270°, is the index or center of the protractor. (Figure 66)

b. When using the protractor, the base line is always oriented parallel to a north-south grid line. The

0° or 360° mark is always toward the top or north on the map and the 90° mark is to the right.

(1) To determine the grid azimuth—

(a) Draw a line connecting the two points (A and B).

(b) Place the index of the protractor at the point where the drawn line crosses a vertical (north-south) grid line.

(c) Keeping the index at this point, align the 0° to 180° line of the protractor on the vertical grid line.

(d) Read the value of the angle from the scale; this is the grid azimuth from point A to point B (Figure 64).

(2) To plot an azimuth from a known point on a map (Figure 6-7)—

(a) Convert the azimuth from magnetic to grid, if necessary. (See paragraph 66.)

(b) Place the protractor on the map with the index mark at the center of mass of the known point and the base line parallel to a north-south grid line.

SEMICIRCULAR

CIRCULAR

RECTANGULAR

SQUARE

Figure 6-5. Types of protractors.

Figure 6-6. Military protractor.

Figure 6-7. Plotting an azimuth on the map.

(c) Make a mark on the map at the desired azimuth.

(d) Remove the protractor and draw a line connecting the known point and the mark on the map. This is the grid direction line (azimuth).

NOTE: When measuring an azimuth, the reading is always to the nearest degree or 10 mils. Distance does not change an accurately measured azimuth.

c. To obtain an accurate reading with the protractor (to the nearest degree or 10 mile), there are two techniques to check that the base line of the protractor is parallel to a north-south grid line.

(1) Place the protractor index where the azimuth line cuts the north-south grid line, aligning the base line of the protractor directly over the intersection of the azimuth line with the north-south grid line. The user should be able to determine whether the initial azimuth reading was correct.

(2) The user should re-read the azimuth between the azimuth and north-south grid line to check the initial azimuth.

(3) Note that the protractor is cut at both the top and bottom by the same north-south grid line. Count the number of degrees from the 0° mark at the top of the protractor to this north-south grid line and then count the number of degrees from the 180° mark at the bottom of the protractor to the same grid line. If the two counts are equal, the protractor is properly aligned.

. DECLINATION DIAGRAM

clination is the angular difference between any two
ths. If you have a map and a compass, the one of
st interest to you will be between magnetic and grid
th. The declination diagram (Figure 6-8) shows the
gular relationship, represented by prongs, among
d, magnetic, and true norths. While the relative
itions of the prongs are correct, they are seldom
tted to scale. Do not use the diagram to measure a
nerical value. This value will be written in the map
rgin (in both degrees and mils) beside the diagram.

a. **Location.** A declination diagram is a part of the
ormation in the lower margin on most larger maps.
medium-scale maps, the declination information is
own by a note in the map margin.

b. **The Grid-Magnetic Angle.** The G-M angle
ue is the angular size that exists between grid north
d magnetic north. It is an arc, indicated by a dashed
e, that connects the grid-north and magnetic-north
ngs. This value is expressed to the nearest 1/2
gree, with mil equivalents shown to the nearest 10
s. The G-M angle is important to the map
der/land navigator because azimuths translated
ween map and ground will be in error by she size of
declination angle if not adjusted for it.

c. **Grid Convergence.** An arc indicated by a
hed line connects the prongs for true north and grid
th. The value of the angle for the center of the sheet
given to the nearest full minute with its equivalent to
nearest milt These data are shown in the form of a
d-convergence note.

d. **Conversion**. There is an angular difference between the grid north and the magnetic north. Since the
location of magnetic north does not correspond exactly
with the grid-north lines on the maps, a conversion from
magnetic to grid or vice versa is needed.

(1) *With notes*. Simply refer to the conversion notes
that appear in conjunction with the diagram explaining the
use of the G-M angle (Figure 6-8). One note provides
instructions for converting magnetic azimuth to grid
azimuth; the other, for converting grid azimuth to
magnetic azimuth. The conversion (add or subtract) is
governed by the direction of the magnetic-north prong
relative to that of the north-grid prong.

(2) *Without notes*. In some cases, there are no
declination conversion notes on the margin of the map; it
is necessary to convert from one type of declination to
another. A magnetic compass gives a magnetic azimuth;
but in order to plot this line on a "ridded map, the
magnetic azimuth value must be changed to grid azimuth.
The declination diagram is used for these conversions. A
rule to remember when solving such problems is this: **No
matter where the azimuth line points, the angle to it is
always measured clockwise from the reference
direction (base line). With this in mind, the problem Is
solved by the following steps:**

(a) Draw a vertical or grid-north line (prong). Always
align this line with the vertical lines on a map (Figure
6-9).

Figure 6-8. Declination diagrams.

(b) From the base of the grid-north line (prong), draw an arbitrary line (or any azimuth line) at a roughly right angle to north, regardless of the actual value of the azimuth in degrees (Figure 6-9).

(c) Examine the declination diagram on the map and determine the direction of the magnetic north (right-left or east-west) relative to that of the grid-north prong. Draw a magnetic prong from the apex of the grid-north line in the desired direction (Figure 6-9).

(d) Determine the value of the G-M angle. Draw an arc from the grid prong to the magnetic prong and place the value of the GM angle (Figure 6-9).

(e) Complete the diagram by drawing an arc from each reference line to the arbitrary line. A glance at the completed diagram shows whether the given azimuth or the desired azimuth is greater, and thus whether the known difference between the two must be added or subtracted.

(f) The inclusion of the true-north prong in relationship to the conversion is of little importance.

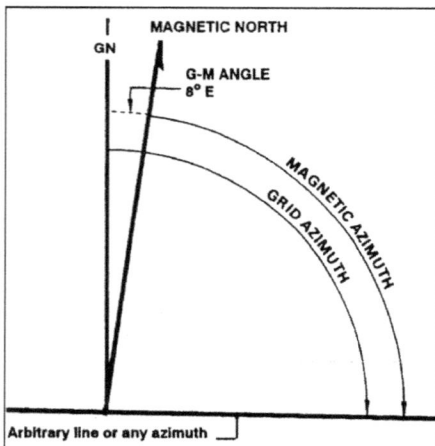

Figure 6-9. Declination diagram with arbitrary line

e. **Applications.** Remember, there are no negative azimuths on the azimuth circle. Since 0° is the same as 360°, then 2° is the same as 362°. This is because 2° and 362° are located at the same point on the azimuth circle. The grid azimuth can now be converted into a magnetic azimuth because the grid azimuth is now larger than the G-M angle.

(1) When working with a map having an east G-M angle:

(a) To plot a magnetic azimuth on a map, fir change it to a grid azimuth (Figure 6-10).

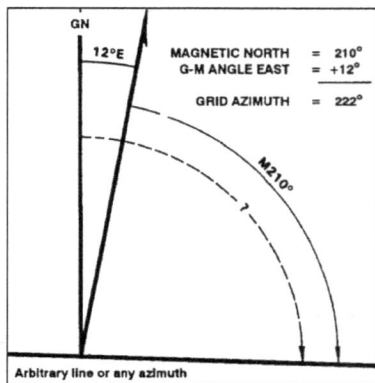

Figure 6-10. Converting to grid azimuth.

(b) To use a magnetic azimuth in the field with compass, first change the grid azimuth plotted on a map t a magnetic azimuth (Figure 6-11).

Figure 6-11. Converting to magnetic azimuth.

(c) Convert a grid azimuth to a magnetic azimuth when the G-M angle is greater than a grid azimuth (Figure 6-12)

Figure 6-12. Converting to a magnetic azimuth when the G-M angle is greater.

(2) When working with a map having a west G-M angle:

(a) To plot a magnetic azimuth on a map, first convert it to a grid azimuth (Figure 6-13).

Figure 6-13. Converting to a grid azimuth on a map

(b) To use a magnetic azimuth in the field with a compass, change the grid azimuth plotted on a map to a magnetic azimuth (Figure 6-14).

Figure 6-14. Converting to a magnetic azimuth on a map.

(c) Convert a magnetic azimuth when the G-M angle is greater than the magnetic azimuth (Figure 6-15).

Figure 6-15. Converting to a grid azimuth when the G-M angle is greater.

(3) The G-M angle diagram should be constructed and used each time the conversion of azimuth is required. Such procedure is important when working with a map for the first time. It also may be convenient to construct a G-M angle conversion table on the margin of the map.

NOTE: When converting azimuths, exercise extreme care when adding and subtracting the G-M angle. A simple mistake of 1° could be significant in the field.

6-7. INTERSECT ION

Intersection is the location of an unknown point by successively occupying at least two (preferably three) known positions on the ground and then map sighting on the unknown location. It is used to locate distant or inaccessible points or objects such as enemy targets and danger areas. There are two methods of intersection: the map and compass method and the straightedge method (Figures 6-16 and 6-17).

a. When using the map and compass method—

(1) Orient the map using the compass.

(2) Locate and mark your position on the map.

(3) Determine the magnetic azimuth to the unknown position using the compass.

(4) Convert the magnetic azimuth to grid azimuth.

(5) Draw a line on the map from your position on the grid azimuth.

(6) Move to a second known point and repeat steps 2, 3, 4, and S.

(7) The location of the unknown position is where the lines cross on the map. Determine the grid coordinates to the desired accuracy (Figure 616).

Figure 6-16. Intersection, using map and compass.

b. The straightedge method is used when a compass is not available. When using it—

(1) Orient the map on a flat surface by the terrain association method.

(2) Locate and mark your position on the map.

(3) Lay a straightedge on the map with one end at the user's position (A) as a pivot point; rotate the straightedge until the unknown point is sighted along the edge.

(4) Draw a line along the straightedge.

(5) Repeat the above steps at position (B) and check for accuracy.

(6) The intersection of the lines on the map is the location of the unknown point (C). Determine the grid coordinates to the desired accuracy (Figure 6-17).

Figure 6-17. Intersection, using a straightedge.

6-8. RESECTION

Resection is the method of locating one's position on a map by determining the grid azimuth to at least two well-defined locations that can be pinpointed on the map. For greater accuracy, the desired method of resection would be to use three or more well-defined locations.

a. When using the map and compass method (Figure 6-18)—

(1) Orient the map using the compass.

(2) Identify two or three known distant locations on the ground and mark them on the map.

(3) Measure the magnetic azimuth to one of the known positions from your location using a compass.

(4) Convert the magnetic azimuth to a grid azimuth.

(5) Convert the grid azimuth to a back azimuth. Using a protractor, draw a line for the back azimuth on the map from the known position back toward your unknown position.

(6) Repeat 3, 4, and 5 for a second position and a third position, if desired.

(7) The intersection of the lines is your location. Determine the grid coordinates to the desired accuracy.

b. When using the straightedge method—

(1) Orient the map on a flat surface by the terrain association method. (Page 612, Figure 6-19)

(2) Locate at least two known distant locations or prominent features on the ground and mark them on the map.

(3) Lay a straightedge on the map using a known position as a pivot point. Rotate the straightedge until the known position on the map is aligned with the known position on the ground.

(4) Draw a line along the straightedge away from the known position on the ground toward your position.

1. From your unknown location to hilltop 408 magnetic azimuth
$312° + 5° E = 317° G - 180° = 137°$ back azimuth
2. From your unknown location to control tower magnetic azimuth
$13° + 5° E = 18°G + 180° = 198°$ back azimuth

Figure 6-18. Resection with map and compass.

Figure 6-19. Resection with straightedge.

(5) Repeat 3 and 4 using a second known position.

(6) The intersection of the lines on the map is your ation. Determine the grid coordinates to the desired uracy.

. MODIFIED RESECTION

dified resection is the method of locating one's posi- on the map when the person is located on a linear ure on the ground, such as a road, canal, or stream gure 6-20). Proceed as follows:

a. Orient the map using a compass or by terrain ociation.

b. Find a distant point that can be identified on the und and on the map.

c. Determine the magnetic azimuth from your ation to the distant known point.

Convert the magnetic azimuth to a grid azimuth.

e. Convert the grid azimuth to a back azimuth.

ng a protractor, draw a line for the back azimuth on map from the known position back toward your

unknown position.

f. The location of the user is where the line crosses the linear feature. Determine the grid coordinates to the desired accuracy.

6-10. POLAR COORDINATES

A method of locating or plotting an unknown position from a known point by giving a direction and a distance along that direction line is called polar coordinates. The following elements must be present when using polar coordinates (Figure 621).

· Present known location on the map.

· Azimuth (grid or magnetic).

· Distance (in meters).

The use of the laser range finder to determine the range will greatly enhance your accuracy in determining the unknown position's location.

Figure 6-20. Modified resection.

1. Your unknown location is along Prairie Creek.
2. From your unknown location to the Church of God, the magnetic azimuth is 20° + 7° E = 27° G + 180° = 207° back azimuth.

Known location — WATER TANK
Grid azimuth — 24°
Distance — 3,600 Meters

Figure 6-21. Polar plot.

CHAPTER 7

OVERLAYS

An overlay is a clear sheet of plastic or semitransparent paper. It is used to display supplemental map and tactical information related to military operations. It is often used as a supplement to orders given in the field. Information is plotted on the overlay at the same scale as on the map, aerial photograph, or other graphic being used. When the overlay is placed over the graphic, the details plotted on the overlay are shown in their true position.

7-1. PURPOSE
Overlays are used to display military operations with enemy and friendly troop dispositions, and as supplements to orders sent to the field. They show detail that will aid in understanding the orders, displays of communication networks, and so forth. They are also used as annexes to reports made in the field because they can clarify matters that are difficult to explain clearly in writing.

7-2. MAP OVERLAY
There are three steps in the making of a map overlay—orienting the overlay material, plotting and symbolizing the detail, and adding the required marginal information (Figure 7-1).

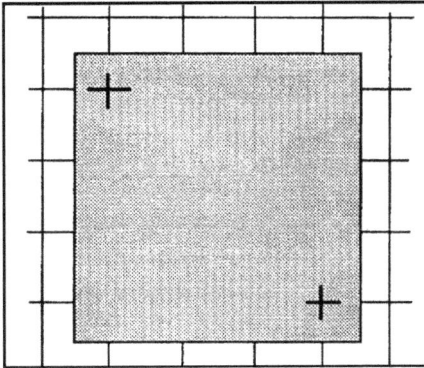

Figure 7-1. Registering the overlay.

a. **Orienting.** Orient the overlay over the place on the map to be annotated. Then, if possible, attach it to the edges of the map with tape. Trace the grid intersections nearest the two opposite corners of the overlay using a straightedge and label each with the proper grid coordinates. These register marks show the receiver of your overlay exactly where it fits on his map; without them, the overlay is difficult to orient. It is imperative that absolute accuracy be maintained in plotting the register marks, as the smallest mistake will throw off the overlay.

b. **Plotting of New Detail.** Use pencils or markers in standard colors that make a lasting mark without cutting the overlay to plot any detail (FM 101-5-1).

(1) Use standard topographic or military symbols where possible. Nonstandard symbols invented by the author must be identified in a legend on the overlay. Depending on the conditions under which the overlay is made, it may be advisable to plot the positions first on the map, then trace them onto the overlay. Since the overlay is to be used as a supplement to orders or reports and the recipient will have an identical map, show only that detail with which the report is directly concerned.

(2) If you have observed any topographic or cultural features that are not shown on the map, such as a new road or a destroyed bridge, plot their positions as accurately as possible on the overlay and mark with the standard topographic symbol.

(3) If difficulty in seeing through the overlay material is encountered while plotting or tracing detail, lift the overlay from time to time to check orientation of information being added in reference to the base.

c. **Recording Marginal Information.** When all required detail has been plotted or traced on the overlay, print information as close to the lower right-hand corner as detail permits (Figure 7-2). This information includes the following data:

(1) *Title and objective.* This tells the reader why the overlay was made and may also give the actual location. For example, "Road Reconnaissance" is not as specific as "Route 146 Road Reconnaissance."

(2) *Time and date.* Any overlay should contain the latest possible information. An overlay received in time

7-1

is very valuable to the planning staff and may affect the entire situation; an overlay that has been delayed for any reason may be of little use. Therefore, the exact time the information was obtained aids the receivers in determining its reliability and usefulness.

(3) *Map reference.* The sheet name, sheet number, map series number, and scale must be included. If the reader does not have the map used for the overlay, this provides the information necessary to obtain it.

(4) *Author.* The name, rank, and organization of the author, supplemented with a date and time of preparation of the overlay, tells the reader if there was a time

difference between when the information was obtained and when it was reported.

(5) *Legend.* If it is necessary to invent nonstandard symbols to show the required information, the legend must show what these symbols mean.

(6) *Security classification.* This must correspond to the highest classification of either the map or the information placed on the overlay. If the information and map are unclassified, this will be so stated. The locations of the classification notes are shown in Figure 7-2, and the notes will appear in both locations as shown.

Figure 7-2. Map overlay with marginal information.

(7) *Additional information*. Any other information that amplifies the overlay will also be included. Make it as brief as possible.

7-3. AERIAL PHOTOGRAPH OVERLAY

Overlays of single aerial photographs are constructed and used in the same way as map overlays. The steps followed are essentially the same, with the following exceptions:

a. **Orienting of Overlay.** The photograph normally does not have grid lines to be used as register marks. The borders of the photograph limit the area of the overlay, so the reference marks or linear features are traced in place of grid register marks. Finally, to ensure proper

location of the overlay with respect to the photograph, indicate on the overlay the position of the marginal data on the photograph as seen through the overlay.

b. **Marginal Information.** The marginal information shown on photographs varies somewhat from that shown on maps. Overlays of photographs (Figure 7-3) should show the following information:

(1) *North arrow*. This may be obtained in two ways— by comparing with a map of the area or by orienting the photograph by inspection. In the latter case, a compass or expedient direction finder must be used to place the direction arrow on the overlay. Use

Figure 7-3. Photographic overlay with marginal information.

the standard symbol to represent the actual north arrow used—grid, magnetic, or true north.

(2) *Title and objective*. This tells the reader why the photo overlay was made and may also give the actual location.

(3) *Time and date*. The exact time the information was obtained is shown on a photo overlay just as on a map overlay.

(4) *Photo reference*. The photo number, mission number, date of flight, and scale appear here, or the information is traced in its actual location on the photograph.

(5) *Scale*. The scale must be computed since it is not part of the marginal data.

(6) *Map reference*. Reference is made to the sheet name, sheet number, series number, and scale of a map of the area, if one is available.

(7) *Author*. The name, rank, and organization of the author are shown, supplemented with a date and time of preparation of the overlay.

(8) *Legend*. As with map overlays, this is only necessary when nonstandard symbols are used.

(9) *Security classification*. This must correspond to the highest classification of either the photograph or the information placed on the overlay. If the information and photograph are unclassified, this will be so stated. The locations of the classification notes are shown in Figure 7-3, and the notes will appear in both locations.

(10) *Additional information*. Any other information that amplifies the overlay will also be included. Make it as brief as possible.

CHAPTER 8

AERIAL PHOTOGRAPHS

An aerial photograph is any photograph taken from an airborne vehicle (aircraft, drones, balloons) satellites, and so forth). The aerial photograph has many uses in military operations; however, for the purpose of this manual, it will be considered primarily as a map supplement or map substitute.

8-1. COMPARISON WITH MAPS

A topographic map may be obsolete because it was compiled many years ago. A recent aerial photograph shows any changes that have taken place since the map was made. For this reason, maps and aerial photographs complement each other. More information can be gained by using the two together than by using either alone.

a. **Advantages.** An aerial photograph has the following advantages over a map:

(1) It provides a current pictorial view of the ground that no map can equal.

(2) It is more readily obtained. The photograph may be in the hands of the user within a few hours after it is taken; a map may take months to prepare.

(3) It may be made for places that are inaccessible to ground soldiers.

(4) It shows military features that do not appear on maps.

(5) It can provide a day-to-day comparison of selected areas, permitting evaluations to be made of enemy activity.

(6) It provides a permanent and objective record of the day-to-day changes with the area.

b. **Disadvantages.** The aerial photograph has the following disadvantages as compared to a map:

(1) Ground features are difficult to identify or interpret without symbols and are often obscured by other ground detail as, for example, buildings in wooded areas.

(2) Position location and scale are only approximate.

(3) Detailed variations in the terrain features are not readily apparent without overlapping photography and a stereoscopic viewing instrument.

(4) Because of a lack of contrasting colors and tone, a photograph is difficult to use in poor light.

(5) It lacks marginal data.

(6) It requires more training to interpret than a map.

8-2. TYPES

Aerial photography most commonly used by military personnel may be divided into two mayor types, the vertical and the oblique. Each type depends upon the attitude of the camera with respect to the earth's surface when the photograph is taken.

a. Vertical. A vertical photograph is taken with the camera pointed as straight down as possible (Figures 8-1 and 8-2). Allowable tolerance is usually + 3° from the perpendicular (plumb) line to the camera axis. The result is coincident with the camera axis. A vertical photograph has the following characteristics:

(1) The lens axis is perpendicular to the surface of the earth.

(2) It covers a relatively small area.

(3) The shape of the ground area covered on a single vertical photo closely approximates a square or rectangle.

(4) Being a view from above, it gives an unfamiliar view of the ground.

(5) Distance and directions may approach the accuracy of maps if taken over flat terrain.

(6) Relief is not readily apparent.

Figure 8-1. Relationship of the vertical aerial photograph with the ground.

Figure 8-2. Vertical photograph.

Figure 8-3. Relationship of low oblique photograph to the ground.

b. **Low Oblique**. This is a photograph taken with the camera inclined about 30° from the vertical (Figure 8-3 and 8-4.) It is used to study an area before an attack, to substitute for a reconnaissance, to substitute for a map, or to supplement a map. A low oblique has the following characteristics:

(1)   It covers a relatively small area.

(2)   The ground area covered is a trapezoid, although the photo is square or rectangular.

(3)   The objects have a more familiar view, comparable to viewing from the top of a high hill or tall building.

(4)   No scale is applicable to the entire photograph, and distance cannot be measured. Parallel lines on the ground are not parallel on this photograph; therefore, direction (azimuth) cannot be measured.

(5)   Relief is discernible but distorted.

(6)   It does not show the horizon.

Figure 8-4. Low oblique photograph.

c. **High Oblique.** The high oblique is a photograph taken with the camera inclined about 60° from the vertical (Figures 8-5 and 8-6). It has a limited military application; it is used primarily in the making of aeronautical charts. However, it may be the only photography available. A high oblique has the following characteristics:

(1) It covers a very large area (not all usable).

(2) The ground area covered is a trapezoid, but the photograph is square or rectangular.

(3) The view vanes from the very familiar to unfamiliar, depending on the height at which the photograph is taken.

(4) Distances and directions are not measured on this photograph for the same reasons that they are not measured on the low oblique.

(5) Relief may be quite discernible but distorted as in any oblique view. The relief is not apparent in a high altitude, high oblique.

(6) The horizon is always visible.

Figure 8-5. Relationship of high oblique photograph to the ground.

Figure 8-6. High oblique photograph.

d. **Trimetrogon.** This is an assemblage of three photographs taken at the same time, one vertical and two high obliques, in a direction at right angle to the line of flight. The obliques, taken at an angle of 61° from the vertical, sidelap the vertical photography, producing composites from horizon to horizon (Figure 8-7).

Figure 8-7. Relationship of cameras to ground for
trimetrogon photography (three cameras).

e. **Multiple Lens Photography.** These are composite photographs taken with one camera having two or more lenses, or by two or more cameras. The photographs are combinations of two, four, or eight obliques around a vertical. The obliques are rectified to permit assembly as verticals on a common plane.

f. **Convergent Photography.** These are done with a single twin-lens, wide-angle camera, or with two single-lens, wide-angle cameras coupled rigidly in the same mount so that each camera axis converges when intentionally tilted a prescribed amount (usually 15 or 20°)from the vertical. Again, the cameras are exposed at the same time. For precision mapping, the optical axes of the cameras are parallel to the line of flight, and for reconnaissance photography, the camera axes are at high angles to the line of flight.

g. **Panoramic.** The development and increasing use of panoramic photography in aerial reconnaissance has resulted from the need to cover in greater detail more and more areas of the world.

(1) To cover the large areas involved, and to resolve the desired ground detail, present-day recon- naissance systems must operate at extremely high resolution levels. Unfortunately, high-resolution levels and wide-angular coverage are basically contradicting requirements.

(2) A panoramic camera is a scanning type of camera that sweeps the terrain of interest from side to side across the direction of flight. This permits the panoramic camera to record a much wider area of ground than either frame or strip cameras. As in the case of the frame cameras, continuous cover is obtained by properly spaced exposures timed to give sufficient overlap between frames. Panoramic cameras are most advantageous for applications requiring the resolution of small ground detail from high altitudes.

8-3. TYPES OF FILM

Types of film generally used in aerial photography include panchromatic, infrared, and color. Camouflage detection film is also available.

a. **Panchromatic.** This is the same type of film that is used in the average hand-held small camera. It records the amount of light reflected from objects in tones of gray running from white to black. Most aerial photography is taken with panchromatic film.

b. **Infrared.** This is a black-and-white film that is sensitive to infrared waves. It can be used to detect artificial camouflage materials and to take photographs at night if there is a source of infrared radiation.

c. **Color.** This film is the same as that used in the average hand-held camera. It is limited in its use because of the time required to process it and its need for clear, sunny weather.

d. **Camouflage Detection.** This film is a special type that records natural vegetation in a reddish color. When artificial camouflage materials are photographed, they appear bluish or purplish. The name of this film indicates its primary use.

8-4. NUMBERING AND TITLING INFORMATION

Each aerial photograph contains in its margin important information for the photo user. The arrangement, type, and amount of this information is standardized; however, the rapid development of cameras, film, and aeronautical technology since World War II has caused numerous changes in the numbering and titling of aerial photographs. As a result, the photo user may find that the marginal information on older photographs varies somewhat from the standard current practice. With certain camera systems, some of the data are automatically recorded on each exposure, while other systems require that all titling data be added to the film after processing.

a. Standard titling data for aerial photography prepared for the use of the Department of Defense are as follows. For reconnaissance and charting photography, items 2 through 14 and item 19 are lettered on the beginning and end of each roll of film. Items 1 through 9 and item 19 are lettered on each exposure. For surveying and mapping photography, items 2 through 19 are lettered on the beginning and end of each roll of film, and items 1, 2, 3, 5, 6, 7, 8, 9, 13, and 19 are lettered on each exposure.

(1) Negative number.
(2) Camera position.
(3) Taking unit.
(4) Service.
(5) Sortie/mission number.
(6) Date (followed by a double hyphen [=])
(7) Time group and zone letter (GMT).
(8) Focal length.
(9) Altitude.
(10) Kind of photography or imagery.
(11) Geographic coordinates.
(12) Descriptive title.

(13) Project number and or name.
(14) Camera type and serial number.
(15) Cone serial number (if any).
(16) Lens type and serial number.
(17) Magazine type and serial number.
(18) Type of photographic filter used.
(19) Security classification.

b. Automatically recorded data may differ somewhat in arrangement from the sequence listed above, but the same information is available to the photo user. A detailed explanation of the titling items and the codes used to indicate them may be found in TM 5-243.

a. **Comparison Method.** The scale of a vertical aerial photograph is determined by comparing the meas-

8-5. SCALE DETERMINATION

Before a photograph can be used as a map supplement or substitute, it is necessary to know its scale. On a map, the scale is printed as a representative fraction that expresses the ratio of map distance to ground distance,

$RF = \dfrac{MD}{GD}$. On a photograph, the scale is also expressed as a ratio, but is the ratio of the photo distance *(PD)* to

ground distance, $RF = \dfrac{PD}{GD}$. The approximate scale

or average scale *(RF)* of a vertical aerial photograph is determined by either of two methods; the comparison method or the focal length-flight altitude method.

ured distance between two points on the photograph with the measured ground distance between the same two points.

The ground distance is determined by actual measurement on the ground or by the use of the scale on a map of

$$SCALE\ (RF\ ) = \frac{Photo\ Distance}{Ground\ Distance}$$

the same area. The points selected on the photograph must be identifiable on the ground or map of the same area and should be spaced in such a manner that a line connecting them will pass through or nearly through the center of the photograph (Figure 8-8).

b. **Focal Length-Flight Altitude Method.** When the marginal information of a photograph includes the focal length and the flight altitude, the scale of the photo is determined using the following formula (Figure 8-9).

When the ground elevation is at sea level, **H** becomes zero, and the formular is as shown in figure 8-10, page 8-8.

Figure 8-8. Selection of points for scale determination.

Figure 8-9. Computation of scale from terrain level.

Figure 8-10. Basic computation of scale from sea level.

8-6. INDEXING

When aerial photos are taken of an area, it is convenient to have a record of the extent of coverage of each photo. A map on which the area covered by each photo is outlined and numbered or indexed to correspond to the photo is called an index map. There are two methods of preparing index maps.

a. The four-corner method (Figures 8-11 and 8-12) requires location on the map of the exact point corresponding to each corner of the photo. If a recognizable object such as a house or road junction can be found exactly at one of the corners, this point may be used on the map as the corner of the photo. If recognizable objects cannot be found at the corners, then the edges of the photo should be outlined on the map by lining up two or more identifiable objects along each edge; the points where the edges intersect should be the exact corners of the photo. If the photo is not a perfect vertical, the area outlined on the map will not be a perfect square or rectangle. After the four sides are drawn on the map, the number of the photograph is written in the enclosed area for identification. This number should be placed in the same corner as it is on the photo.

Figure 8-11. Four-corner method (selection of points).

POINTS SELECTED

Figure 8-12. Plotting, using the four-corner method.

b. The template method is used when a large number of photos are to be indexed, and the exact area covered by each is not as important as approximate area and location. In this case, a template (cardboard pattern or guide) is cut to fit the average area the photos cover on the index map. It is used to outline the individual area covered by each photo. To construct a template, find the average map dimensions covered by the photos to be indexed as follows. Multiply the average length of the photos by the denominator of the average scale of the photos; multiply this by the scale of the map. Do the same for the width of the photos. This gives the average length and width of the area each photo covers on the map—or the size to which the template should be cut (Figure 8-13, page 8-10).

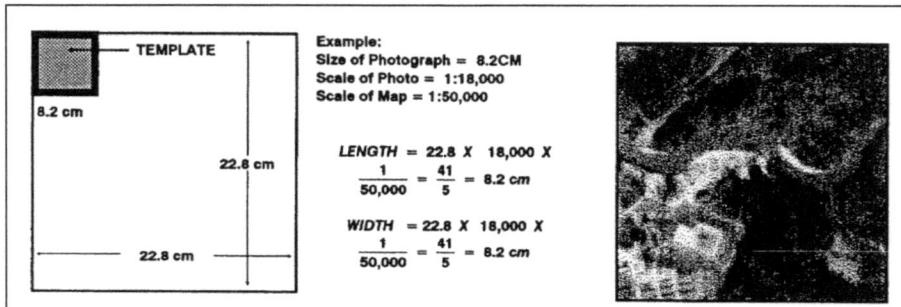

Example:
Size of Photograph = 8.2CM
Scale of Photo = 1:18,000
Scale of Map = 1:50,000

$$LENGTH = 22.8 \; X \; 18,000 \; X \; \frac{1}{50,000} = \frac{41}{5} = 8.2 \; cm$$

$$WIDTH = 22.8 \; X \; 18,000 \; X \; \frac{1}{50,000} = \frac{41}{5} = 8.2 \; cm$$

Figure 8-13. Constructing a template.

c. To index the map, select the general area covered by the first photo and orient the photo to the map. Place the template over the area on the map and adjust it until it covers the area as completely and accurately as possible. Draw lines around the edges of the template. Remove the rectangle and proceed to the next photo (Figure 8-14).

Figure 8-14. Indexing with a template.

d. After all photos have been plotted, write on the map sufficient information to identify the mission or sortie. If more than one sortie is plotted on one map or overlay, use a different color for each sortie.

e. In most cases, when a unit orders aerial photography, an index is included to give the basic information. Instead of being annotated on a map of the area, it appears on an overlay and is keyed to a map.

8-7. ORIENTING OF PHOTOGRAPH

Orienting the photograph is important because it is of very little value as a map supplement or substitute if its location and direction are not known by the user.

a. If a map of the same area as the photograph is available, the photograph is oriented to the map by comparing features common to both and then transferring a direction line from the map to the photograph.

b. If no map is available, the shadows on a photograph may be used to get an approximate true-north line. This method is not recommended in the torrid zone (Figure 8-15).

(1) *North temperate zone*. The sun moves from the east in the morning through south at noon to west in the afternoon. Conversely, shadow fall varies from west through north to east. Before noon, therefore, north is to the right of the direction of shadow fall; at noon, north is the direction of shadow fall; and after noon, north is to the left of shadow fall. On an average, the amount of variation in shadow fall per hour is 15 degrees. From marginal information, determine the number of hours from noon that the photo was taken and multiply that number by 15°. With a protractor, measure an angle of that amount in the proper direction (right to left) from a clear, distinct shadow, and north is obtained. For photographs taken within three hours of noon, a reasonable accurate north direction can be obtained. Beyond these limits, the 15° must be corrected, depending on time of year and latitude.

(2) *South temperate zone*. The sun moves from east through north at noon to west. Shadows then vary from west through south to east. Before noon, south is to the left of shadow fall; at noon, south is shadow fall;

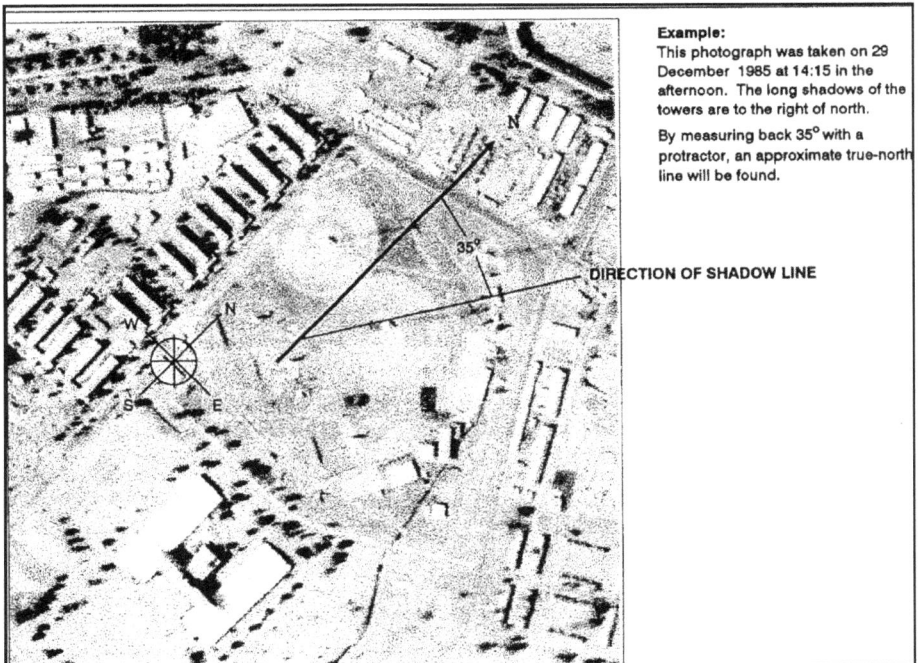

Example:
This photograph was taken on 29 December 1985 at 14:15 in the afternoon. The long shadows of the towers are to the right of north.

By measuring back 35° with a protractor, an approximate true-north line will be found.

Figure 8-15. Using shadows on a photograph to find north.

and after noon, south is to the right of shadow fall. Proceed as in (1) above to determine the direction of south.

c. On a photograph that can be oriented to the surrounding ground features by inspection, a magnetic north line can be established using a compass.

(1) Orient the photograph by inspection.

(2) Open the compass and place it on the photograph.

(3) Without moving the photograph, rotate the compass until the north arrow is under the stationary black line.

(4) Draw a line along the straight edge of the compass. This is a magnetienorth line.

8-8. POINT DESIGNATION GRID

Since aerial photographs are seldom exactly the same scale as a map of the same area, it is not feasible to print military grids on them. A special grid is used for the designation of points on photographs (Figure 8-16). This grid, known as the point designation grid, has no relation to the scale of the photo, to any direction, or to the grid used on any other photograph or map. It has only one purpose, to designate points on photographs.

a. The point designation grid is rarely printed on photographs; therefore, it becomes the responsibility of each user to construct the grid on the photograph. All users must construct the grid in exactly the same way. Before the grid can be constructed or used, the photograph must be held so that the marginal information,

Figure 8-16. Point designation grid.

regardless of where it is located, is in the normal reading position (Figure 8-17, step 1).

(1) Draw lines across the photograph joining opposite reference marks at the center of each photograph (fiducial marks). If there are no fiducial marks, the center of each side of the photograph is assumed to be the location of the marks (Figure 8-17, step 2).

(2) Space grid lines, starting with the center line, 4 centimeters (1.575 inches) apart (a distance equal to 1,000 meters at a scale of 1:25,000). The 1:25,000 map coordinate scale can be used for this dimension and to accurately designate points on the photograph, but this

does not mean that distance can be scaled from the photograph. Extend the grid past the margins of the photograph so that a horizontal and vertical grid line fall outside the picture area (Figure 8-17, step 3).

(3) Number each center line "50" and give numerical values to the remaining horizontal and vertical lines so that they increase to the right and up (Figure 8-17, step 4).

b. The point designation grid is used, once the photograph is oriented, in the same manner as the grid on a map (Figure 8-18), **read right and up**. The coordinate scale used with the UTM grid on maps at the scale of

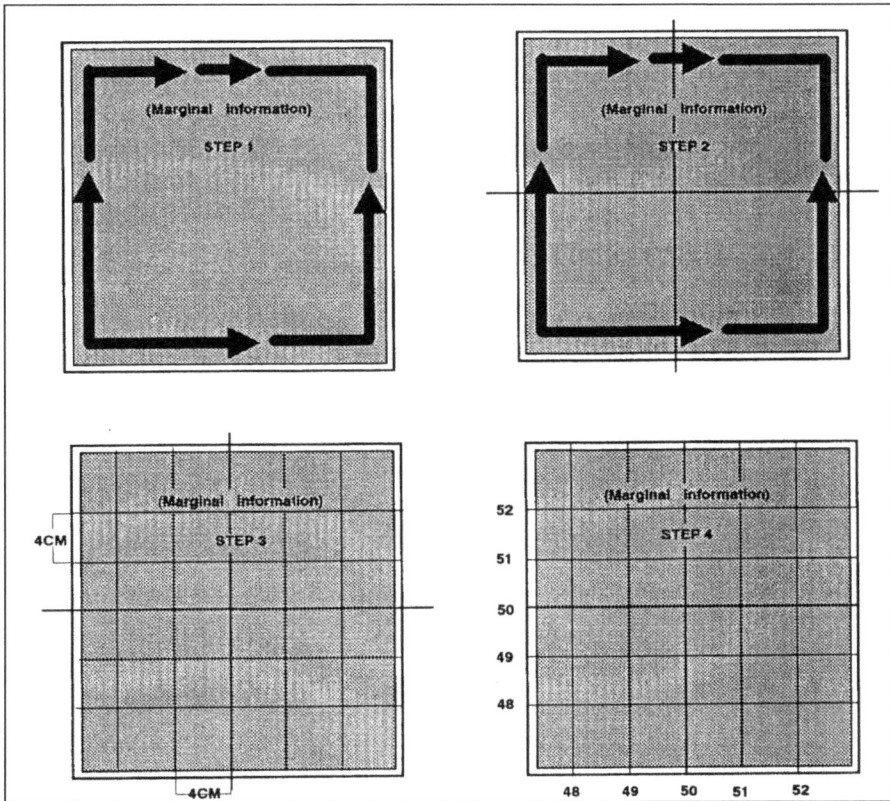

Figure 8-17. Constructing a point designation grid.

1:25,000 may be used to subdivide the grid square in the same manner as on a map. However, because the same point designation grid is used on all photographs, the coordinates of a point on the photograph must be prefixed by the identifying marginal information of the photograph.

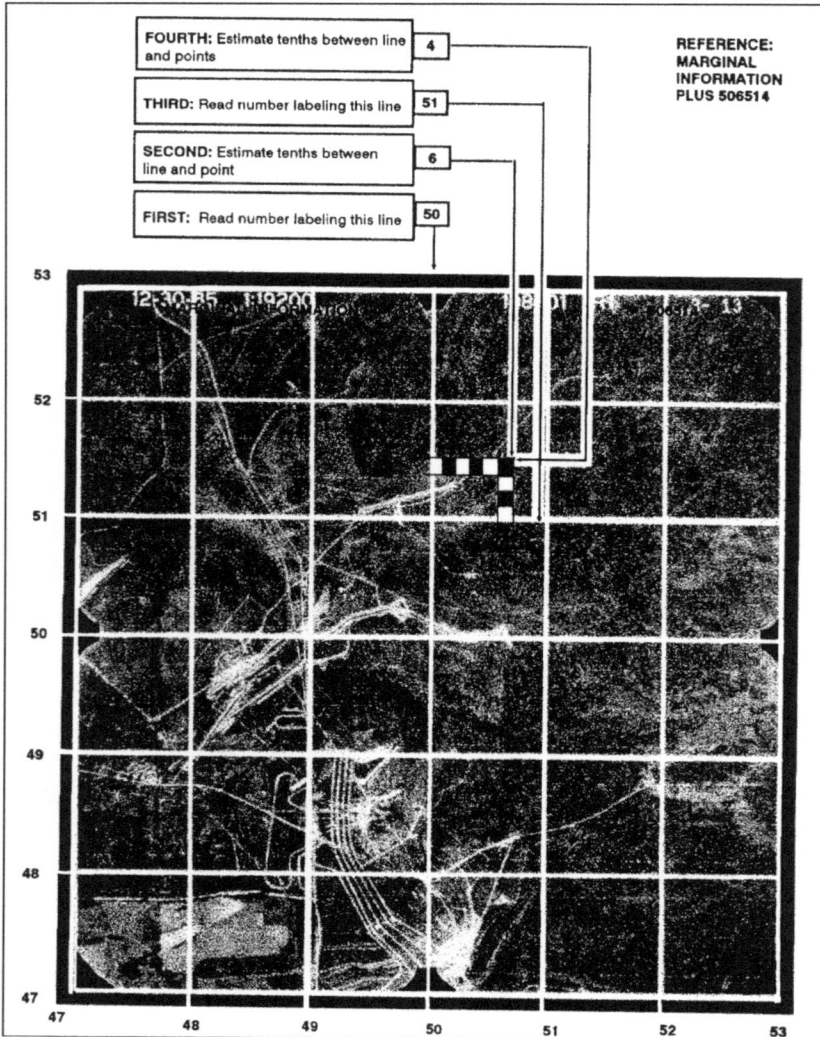

Figure 8-18. Reading point designation grid coordinates

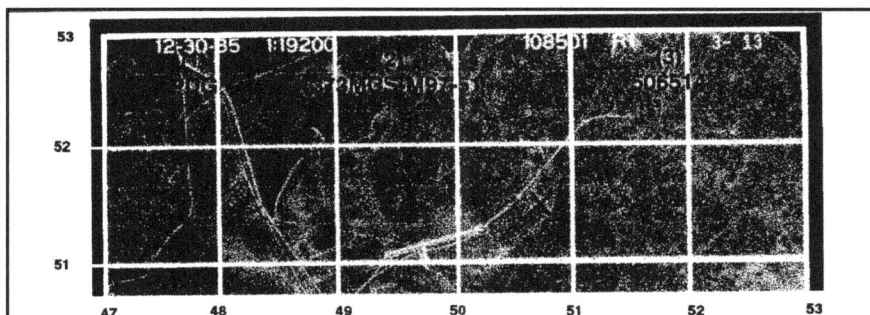

Figure 8-19. Locating the grid coordinate on a point designation grid.

c. A grid coordinate using the point designation grid (Figure 8-19) consists of three parts:

(1) The letters "PDG" to indicate an aerial photograph rather than a map grid coordinate.

(2)The mission and photo negative number to identify which photograph is being used.

(3) The six numerical digits to locate the actual point on the photograph.

8-9. IDENTIFICATION OF PHOTOGRAPH FEATURES

The identification of features on a photograph is not difficult if the following facts are remembered. The view that is presented by the aerial photograph is from above and, as a result, objects do not look familiar. Objects that are greatly reduced in size appear distorted. Most aerial photography is black and white, and all colors appear on the photograph in shades of gray. Generally speaking, the darker the natural color, the darker it will appear on the photograph.

a. The identification of features on aerial photographs depends upon a careful application of five factors of recognition. No one factor will give a positive identification; it requires the use of all five.

(1) *Size*. The size of unknown objects on a photograph, as determined from the scale of the photograph or a comparison with known objects of known size, gives a clue to their identity. For example, in a built-up area the smaller buildings are usually dwellings, and the larger buildings are commercial or community buildings.

(2) *Shape (pattern)*. Many features possess characteristic shapes that readily identify the features. Man-made features appear as straight or smooth curved lines, while natural features usually appear to be irregular. Some of the most prominent man-made features are

highways, railroads, bridges, canals, and buildings. Compare the regular shapes of these to the irregular shapes of such natural features as streams and timber lines.

(3) *Shadows*. Shadows are very helpful in identifying features since they show the familiar side view of the object. Some excellent examples are the shadows of water towers or smoke stacks. As viewed directly from above, only a round circle or dot is seen, whereas the shadow shows the profile and helps to identify the object. Relative lengths of shadows also usually give a good indication of relative heights of objects.

(4) *Shade (tone or texture)*. Of the many different types of photographic film in use today, the film used for most aerial photography, except for special purposes, is panchromatic film. Panchromatic film is sensitive to all the colors of the spectrum; it registers them as shades of gray, ranging from white to black. This lighter or darker shade of features on aerial photographs is known as the tone. The tone is also dependent on the texture of the features; a paved highway has a smooth texture and produces an even tone on the photograph, while a recently plowed field or a marsh has a rough, choppy texture and results in a rough or grainy tone. It is also important to remember that similar features may have different tones on different photographs, depending on the reflection of sunlight. For example, a river or body of water appears light if it is reflecting sunlight directly toward the camera, but appears dark otherwise. Its texture may be smooth or rough, depending on the surface of the water itself. As long as the variables are kept in mind, tone and texture may be used to great advantage.

(5) *Surrounding objects*. Quite often an object not easily recognized by itself may be identified by its

relative position to surrounding objects. Large buildings located beside railroads or railroad sidings are usually factories or warehouses. Schools may be identified by the baseball or football fields. It would be hard to tell the difference between a water tower next to a railroad station and a silo next to a barn, unless the surrounding objects such as the railroad tracks or cultivated fields were considered.

b. Before a vertical photograph can be studied or used for identification of features, it must be oriented. This orienting is different from the orienting required for the construction or use of the point designation grid. Orienting for study consists of rotating the photograph so that the shadows on the photograph point toward yourself. You then face a source of light. This places the source of light, an object, and its shadow in a natural relationship. Failure to orient a photograph properly may cause the height or depth of an object to appear reversed. For example, a mine or quarry may appear to be a hill instead of a depression.

8-10. STEREOVISION

One of the limitations of the vertical aerial photograph is the lack of apparent relief. Stereoscopic vision (or as it is more commonly known, stereovision or depth perception) is the ability to see three-dimensionally or to see length, width, and depth (distance) at the same time. This requires two views of a single object from two slightly different positions. Most people have the ability to see three-dimensionally. Whenever an object is viewed, it is seen twice—once with the left eye and once with the right eye. The fusion or blending together of these two images in the brain permits the judgment of depth or distance.

a. In taking aerial photographs, it is rare for only a single picture to be taken. Generally, the aircraft flies over the area to be photographed taking a series of pictures, each of which overlaps the photograph preceding it and the photograph following it so that an unbroken coverage of the area is obtained (Figure 8-20). The amount of overlap is usually 56 percent, which means that 56 percent of the

Figure 8-20. Photographic overlap.

ground detail appearing on one photo also appears on the next photograph. When a single flight does not give the necessary coverage of an area, additional flights must be made. These additional flights are parallel to the first and must have an overlap between them. This overlap between flights is known as side lap and usually is between 15 and 20 percent (Figure 8-21).

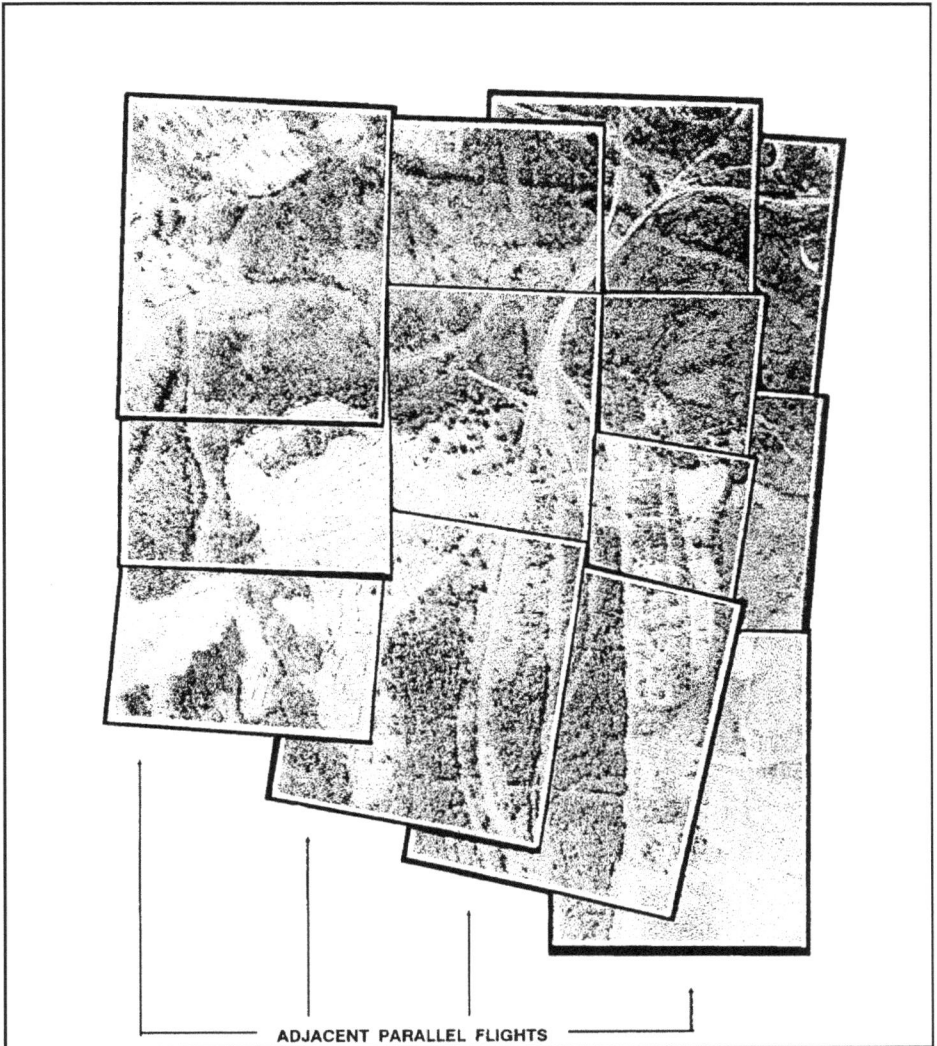

ADJACENT PARALLEL FLIGHTS

Figure 8-21. Side lap.

Figure 8-22. Pocket stereoscope.

Figure 8-23. Mirror stereoscope.

b. The requirement for stereovision can be satisfied by overlapping photographs if one eye sees the object on one photograph and the other eye sees the same object on another photograph. While this can be done after practice with the eyes alone, it is much easier if an optical aid is used. These optical aids are known as stereoscopes. There are many types of stereoscopes, but only the two most commonly used are discussed in this manual.

(1) *Pocket stereoscope*. The pocket stereoscope (Figure 8-22), sometimes known as a lens stereoscope, consists of two magnifying lenses mounted in a metal frame. Because of its simplicity and ease of carrying, it is the type used most frequently by military personnel.

(2) *Mirror stereoscope*. The mirror stereoscope (Figure 8-23) is larger, heavier, and more subject to damage than the pocket stereoscope. It consists of four mirrors mounted in a metal frame.

c. A method to orient a pair of aerial photographs for best three-dimensional viewing is outlined below:

(1) Arrange the selected pair of photos in such a way that the shadows on them generally appear to fal3 toward the viewer. It is also desirable that the light source during the study of the photography enter the side away from the observer (Figure 8-24).

(2) Place the pair of photographs on a flat surface so that the detail on one photograph is directly over the same detail on the other photograph.

(3) Place the stereoscope over the photographs so that the left lens is over the left photograph and the right lens is over the right photograph (Figure 8-24).

(4) Separate the photographs along the line of flight until a piece of detail appearing in the overlap area of the left photograph is directly under the left lens and the same piece of detail on the right photo is directly under the right lens.

(5) With the photograph and stereoscope in this position, a three-dimensional image should be seen. A few minor adjustments may be necessary, such as adjusting the aerial photographs of the stereoscope to obtain the correct position for your eyes. The hills appear to rise and the valleys sink so that

Figure 8-24. Placement of stereoscope over stereopair.

there is the impression of being in an aircraft looking down at the ground.

(6) The identification of features on photographs is much easier and more accurate with this three-dimensional view. The same five factors of recognition (size, shape, shadow, tone, and surrounding objects) must still be applied, but now, with the addition of relief, a more natural view is seen.

PART TWO
LAND NAVIGATION
CHAPTER 9
NAVIGATION EQUIPMENT AND METHODS

Compasses are the primary navigation tools to use when moving in an outdoor world where there is no other way to find directions. Soldiers should be thoroughly familiar with the compass and its uses. Part One of this manual discussed the techniques of map reading To complement these techniques, a mastery of field movement techniques is essential. This chapter describes the lensatic compass end its uses, and some of the field expedient methods used to find directions when compasses are not available.

9-1. TYPES OF COMPASSES

The **lensatic compass** is the most common and simplest instrument for measuring direction. It is discussed in detail in paragraph 9-2. The **artillery M-2 compass** is a special-purpose instrument designed for accuracy; it will be discussed in Appendix G. The **wrist/pocket compass** is a small magnetic compass that can be attached to a wristwatch band. It contains a north-seeking arrow and a dial in degrees. A **protractor** can be used to determine azimuths when a compass is not available. However, it should be noted that when using the protractor on a map, only grid azimuths are obtained.

9-2. LENSATIC COMPASS

The lensatic compass (Figure 9-1, page 9-2) consists of three major parts: the cover, the base, and the lens.

a. **Cover.** The compass cover protects the floating dial. It contains the sighting wire (front sight) and two luminous sighting slots or dots used for night navigation.

b. **Base.** The body of the compass contains the following movable parts:

(1) The floating dial is mounted on a pivot so it can rotate freely when the compass is held level. Printed on the dial in luminous figures are an arrow and the letters E and W. The arrow always points to magnetic north and the letters fall at east (E) 90° and west (W) 270° on the dial. There are two scales; the outer scale denotes mils and the inner scale (normally in red) denotes degrees.

(2) Encasing the floating dial is a glass containing a fixed black index line.

(3) The bezel ring is a ratchet device that clicks when turned. It contains 120 clicks when rotated fully; each click is equal to 3°. A short luminous line that is used in conjunction with the north-seeking arrow during navigation is contained in the glass face of the bezel ring.

(4) The thumb loop is attached to the base of the compass.

c. **Lens.** The lens is used to read the dial, and it contains the rear-sight slot used in conjunction with the front fo sighting on objects. The rear sight also serves as a lock and clamps the dial when closed for its protection. The rear sigh must he opened more than 45° to allow the dial to floa freely.

NOTE: When opened, the straightedge on the left side o the compass has a coordinate scale; the scale is 1:50,000 in newer compasses.

> **WARNING**
> Some older compasses will have a 1:25,000 scale. This scale can be used with a 1:50,000-scale map, but the values read must be halved. Check the scale.

9-3. COMPASS HANDLING

Compasses are delicate instruments and should be cared fo accordingly.

a. **Inspection.** A detailed inspection is required when first obtaining and using a compass. One of the most important parts to check is the floating dial, which contains the magnetic needle. The user must also make sure the sighting wire is straight, the glass and crystal parts are not broken, the numbers on the dial are readable, and most important, that the dial does not stick.

b. **Effects of Metal and Electricity.** Metal objects and electrical sources can affect the performance of a compass. However , nonmagnetic metals and alloys do not affect compass readings. The following separation

Figure 9-1. Lensatic compass.

istances are suggested to ensure proper functioning of compass:

High-tension power lines ....55 meters.
Field gun, truck, or tank ....18 meters.
Telegraph or telephone wires
and barbed wire ....10 me t e rs.
Machine gun ....2 meters.
Steel helmet or rifle ....1/2 meter.

c. **Accuracy.** A compass in good working condition very accurate. However, a compass has to be checked eriodically on a known line of direction, such as a rveyed azimuth using a declination station. Compasses ith more than 3° + variation should not be used.

d. **Protection.** If traveling with the compass nfolded, make sure the rear sight is fully folded down nto the bezel ring. This will lock the floating dial and revent vibration, as well as protect the crystal and rear ght from damage.

-4. USING A COMPASS

lagnetic azimuths are determined with the use of lagnetic instruments, such as lensatic and M-2 ompasses. The techniques employed when using the nsatic compass are as follows:

a. **Using the Centerhold Technique.** First, open the compass to its fullest so that the cover forms a straightedge with the base. Move the lens (rear sight) to the rearmost position, allowing the dial to float freely. Next, place your thumb through the thumb loop, form a steady base with your third and fourth fingers, and extend your index finger along the side of the compass. Place the thumb of the other hand between the lens (rear sight) and the bezel ring; extend the index finger along the remaining side of the compass, and the remaining fingers around the fingers of the other hand. Pull your elbows firmly into your sides; this will place the compass between your chin and your belt. To measure an azimuth, simply turn your entire body toward the object, pointing the compass cover directly at the object. Once you are pointing at the object, look down and read the azimuth from beneath the fixed black index line (Figure 9-2). This preferred method offers the following advantages over the sighting technique:

(1) It is faster and easier to use.
(2) It can be used under all conditions of visibility.
(3) It can be used when navigating over any type of terrain.
(4) It can be used without putting down the rifle; however, the rifle must he slung well back over either shoulder.
(5) It can be used without removing eyeglasses.

Figure 9-2. Centerhold technique.

b. **Using the Compass-to-Cheek Technique.** Fold the cover of the compass containing the sighting wire to a vertical position; then fold the rear sight slightly forward. Look through the rear-sight slot and align the front-sight hairline with the desired object in the distance. Then glance down at the dial through the eye lens to read the azimuth (Figure 9-3).

NOTE: The compass-to-cheek technique is used almost exclusively for sighting, and it is the best technique for this purpose.

Figure 9-3. Compass-to-cheek technique.

c. **Presetting a Compass and Following an azimuth.** Although different models of the lensatic mpass vary somewhat in the details of their use, the nciples are the same.

(1) During daylight hours or with a light source:

(a) Hold the compass level in the palm of the hand.

(b) Rotate it until the desired azimuth falls under the ed black index line (for example, 320°), maintaining e azimuth as prescribed (Figure 9-4).

(c) Turn the bezel ring until the luminous line is gned with the north-seeking arrow. Once the gnment is obtained, the compass is preset.

(d) To follow an azimuth, assume the centerhold :hnique and turn your body until the north-seeking row is aligned with the luminous line. Then proceed rward in the direction of the front cover's sighting re, which is aligned with the fixed black index line at contains the desired azimuth.

(2) During limited visibility, an azimuth may be set the compass by the click method. Remember that the zel ring contains 3° intervals (clicks).

(a) Rotate the bezel ring until the luminous line is er the fixed black index line.

(b) Find the desired azimuth and divide it by three. The result is the number of clicks that you have to rotate the bezel ring.

(c) Count the desired number of clicks. If the desired azimuth is smaller than 180°, the number of clicks on the bezel ring should be counted in a counterclockwise direction. For example, the desired azimuth is 51°. Desired azimuth is 51° 3 = 17 clicks counterclockwise. If the desired azimuth is larger than 180°, subtract the number of degrees from 360° and divide by 3 to obtain the number of clicks. Count them in a clockwise direction. For example, the desired azimuth is 330°; 360°- 330°=30 . 3= 10 clicks clockwise.

(d) With the compass preset as described above, assume a centerhold technique and rotate your body until the north-seeking arrow is aligned with the luminous line on the bezel. Then proceed forward in the direction of the front cover's luminous dots, which are aligned with the fixed black index line containing the azimuth.

(e) When the compass is to be used in darkness, an initial azimuth should be set while light is still available, if possible. With the initial azimuth as a

Figure 9-4. Compass preset at 320 degrees.

base, any other azimuth that is a multiple of three can be established through the use of the clicking feature of the bezel ring.

NOTE: Sometimes the desired azimuth is not exactly divisible by three, causing an option of rounding up or rounding down. If the azimuth is rounded up, this causes an increase in the value of the azimuth, and the object is to be found on the left. If the azimuth is rounded down, this causes a decrease in the value of the azimuth, and the object is to be found on the right.

d. **Bypassing an Obstacle.** To bypass enemy positions or obstacles and still stay oriented, detour around the obstacle by moving at right angles for specified distances.

(1) For example, while moving on an azimuth of 90°, change your azimuth to 180° and travel for 100 meters; change your azimuth to 90° and travel for 150 meters; change your azimuth to 360° and travel for 100 meters; then change your azimuth to 90° and you are back on your original azimuth line (Figure 9-5).

(2) Bypassing an unexpected obstacle at night is a fairly simple matter. To make a 90° turn to the right, hold the compass in the centerhold technique; turn until the center of the luminous letter E is under the luminous line (do not move the bezel ring). To make a 90° turn to the left, turn until the center of the luminous letter W is under the luminous line. This does not require changing the compass setting (bezel ring), and it ensures accurate 90° turns.

Figure 9-5. Bypassing an obstacle.

e. **Offset.** A deliberate offset is a planned magnetic
iation to the right or left of an azimuth to an
ective. Use it when the objective is located along or
the vicinity of a linear feature such as a road or
eam. Because of errors in the compass or in map
ding, the linear feature may he reached without
owing whether the objective lies to the right or left.
deliberate offset by a known number of degrees in a
own direction compensates for possible errors and
sures that upon reaching the linear feature, the user
ows whether to go right or left to reach the objective.
n degrees is an adequate offset for most tactical uses.
ch degree offset will move the course about 18
ters to the right or left for each 1,000 meters
veled. For example, in Figure 9-6, the number of
grees offset is 10. If the distance traveled to "x" in
00 meters, then "x" is located about 180 meters to
right of the objective.

5. FIELD-EXPEDIENT METHODS

en a compass is not available, different techniques
uld be used to determine the four cardinal
ections.

Shadow-Tip Method.

(1) This simple and accurate method of finding
ection by the sun consists of four basic steps (Figure
7).

ep 1. Place a stick or branch into the ground at a
el spot where a distinctive shadow will be cast.
rk the shadow tip with a stone, twig, or other
ans. This first shadow mark is always the west
ection.

p 2. Wait 10 to 15 minutes until the shadow tip
ves a few inches. Mark the new position of
shadow tip in the same way as the first.

Step 3. Draw a straight line through the two marks to
obtain an approximate east-west line.

Step 4. Standing with the first mark (west) to your left,
the other directions are simple; north is to the front,
east is to the right, and south is behind you.

(2) A line drawn perpendicular to the east-west
line at any point is the approximate north-south line. If
you are uncertain which direction is east and which is
west, observe this simple rule—the first shadow-tip
mark is always in the west direction, everywhere on
earth.

(3) The shadow-tip method can also be used as a
shadow clock to find the approximate time of day
(Figure 9-7).

(a) To find the time of day, move the stick to the
intersection of the east-west line and the north-south
line, and set it vertically in the ground. The west part
of the east-west line indicates 0600 hours, and the east
part is 1800 hours, anywhere on earth, because the
basic rule always applies.

(b) The north-south line now becomes the noon
line. The shadow of the stick is an hour hand in the
shadow clock, and with it you can estimate the time
using the noon line and the 6 o'clock line as your
guides. Depending on your location and the season, the
shadow may move either clockwise or counterclockwise, but this does not alter your manner of
reading the shadow clock.

(c) The shadow clock is not a timepiece
in the ordinary sense. It makes every day 12
unequal hours long, and always reads 0600
hours at sunrise and 1800 hours at sunset. The
shadow clock time is closest to conventional clock
time at midday, but the spacing of the

Figure 9-6. Deliberate offset to the objective.

Figure 9-7. Determining directions and time by shadow.

other hours compared to conventional time varies some-what with the locality and the date. However, it doe provide a satisfactory means of telling time in the absence of properly set watches.

(d) The shadow-tip system is not intended for us in polar regions, which the Department of Defense defines as, being above 60° latitude in either hemisphere. Distressed persons in these areas are advised to stay in one place so that search/rescue teams may easily find them. The presence and location of all aircraft and ground parties in polar regions are reported to and checked regularly by governmental or other agencies, and any need for help becomes quickly known.

b. **Watch Method.**

(1) A watch can be used to determine the approximate true north and true south. In the north temperate zone only, the hour hand is pointed toward the sun. A south line can be found midway between the hour hand and 1200 hours, standard time. If on daylight saving time, the north-south line is found between the hour hand and 1300 hours. If there is any doubt as to which end of the line is north, remember that the sun is in the east before noon and in the west after noon.

(2) The watch may also be used to determine direction in the south temperate zone; however, the method is different. The 1200-hour dial is pointed toward the sun, and halfway

between 1200 hours and the hour hand will be a north line. on daylight saving time, the north line lies midway betwe the hour hand and 1300 hours(Figure 9-8).

(3)The watch method can be in error, especially the lower latitudes, and may cause circling. To avc this, make a shadow clock; and set your watch to t time indicated- After traveling for an hour, take anoth shadow-clock reading. Reset your watch if necessa

Figure 9-8. Determining direction by using a wat

9-

c. Star Method.

(1) Less than 60 of approximately 5,000 stars visible the eye are used by navigators. The stars seen as we k up at the sky at night are not evenly scattered across whole sky. Instead they are in groups called stellations.

(2) The constellations that we see depend partly on ere we are located on the earth, the time of the year, the time of the night. The night changes with the sons because of the journey of the earth around the , and it also changes from hour to hour because the ning of the earth makes some constellations seem to el in a circle. But there is one star that is in almost ctly the same place in the sky all night long every ht. It is the North Star, also known as the Polar Star Polaris.

(3) The North Star is less than 1° off true north and does not move from its place because the axis of the earth is pointed toward it. The North Star is in the group of stars called the Little Dipper. It is the last star in the handle of the dipper. Two stars in the Big Dipper are a help in finding the North Star. They are called the Pointers, and an imaginary line drawn through them five times their distance points to the North Star. There are many stars brighter than the North Star, but none is more important because of its location. However, the North Star can only be seen in the northern hemisphere so it cannot serve as a guide south of the equator. The farther one goes north, the higher the North Star is in the sly, and above latitude 70°, it is too high in the sky to be useful (Figure 9-9).

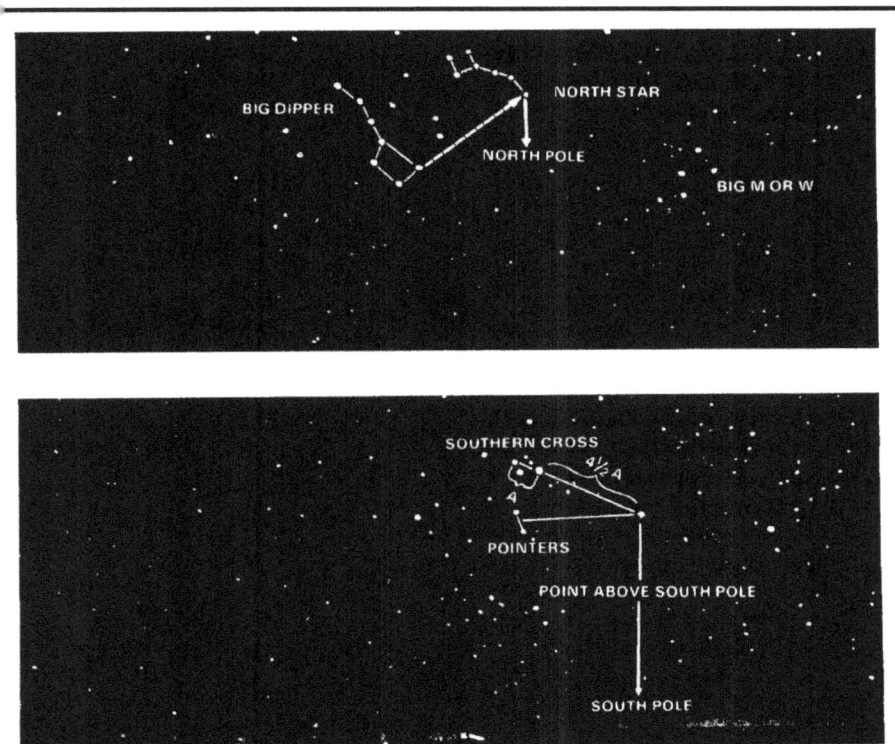

Figure 9-9. Determining direction by the North Star and the Southern Cross.

(4) Depending on the star selected for navigation, azimuth checks are necessary. A star near the north horizon serves for about a half hour. When moving south, azimuth checks should be made every 15 minutes. When traveling east or west, the difficulty of staying on azimuth is caused more by the likelihood of the star climbing too high in the sky or losing itself behind the western horizon than it is by the star changing direction

angle. When this happens, it is necessary to change to another guide star. The Southern Cross is the main constellation used as a guide south of the equator and the above general directions for using north and south stars are reversed. When navigating using the stars as guides, the user must know the different constellation shapes and their locations throughout the world (Figures 9-10 and 9-11).

Figure 9-10. Constellations, Northern Hemisphere.

Figure 9-11. Constellations, Southern Hemisphere.

GLOBAL POSITIONIN& SYSTEM

GPS is a space-based, global, all-weather, tinuously available, radio positioning navigation em. It is highly accurate in determining position tion derived from signal triangulation from a llite constellation system. It is capable of determining ude, longitude, and altitude of the individual user. It being fielded in hand-held, manpack, vehicular, raft, and watercraft configurations. The GPS receives processes datafrom satellites on either a simultaneous or sequential basis. It measures the velocity and range with respect to each satellite; processes the data in terms of an earth centered, earth-fixed coordinate system; and displays the information to the user in geographic or military grid coordinates.

a. The GPS can provide precise steering information, as well as position location. The receiver can accept many checkpoints entered in any coordinate system by the user and convert them to the desired coordinate system. The user then calls up the desired checkpoint and the receiver will display direction and distance to the checkpoint. The GPS does not have inherent drift, an improvement over the Inertial Navigation System, and the receiver will automatically update its position. The receiver can also compute time to the next checkpoint.

c. Specific uses for the GPS are position location; navigation; weapon location; target and sensor location; coordination of firepowe, scout and screening opera-tions; combat resupply; location of obstacles, barriers, and gaps; and communication support. The GPS also has the potential to allow units to train their soldiers and provide the following:

•Performance feedback.

•Knowledge of routes taken by the soldier.

•Knowledge of errors committed by the soldier.

•Comparison of planned versus executed routes.

•Safety and control of lost and injured soldiers. (See Appendix J for more information of the GPS.)

ELEVATION AND RELIEF

The elevation of points on the ground and the relief of an area affect the movement, positioning, and, in some cases, effectiveness of military units. Soldiers must know how to determine locations of points on a map, measure distances and azimuths, and identify symbols on a map. They must also be able to determine the elevation and relief of areas on standard military maps. To do this, they must first understand how the mapmaker indicated the elevation and relief on the map.

10-1. DEFINITIONS

There must be a reference or start point to measure anything. The reference or start point for vertical measurement of elevation on a standard military map is the **datum plane** or **mean sea level,** the point halfway between high tide and low tide. **Elevation** of a point on the earth's surface is the vertical distance it is above or below mean sea level. **Relief** is the representation (as depicted by the mapmaker) of the shapes of hills, valleys, streams, or terrain features on the earth's surface.

10-2. METHODS OF DEPICTING RELIEF

There are several methods used by mapmakers to depict relief of the terrain.

a. **Layer Tinting.** Layer tinting is a method of showing relief by color. A different color is used for each band of elevation. Each shade of color, or band, represents a definite elevation range A legend is printed on the map margin to indicate the elevation range represented by each color. However, this method does not allow the map user to determine the exact elevation of a specific point — only the range.

b. **Form Lines.** Form lines are not measured from any datum plane. Form lines have no standard elevation and give only a general idea of relief. Form lines are represented on a map as dashed lines and are never labeled with representative elevations.

c. **Shaded Relief.** Relief shading indicates relief by a shadow effect achieved by tone and color-that results in the darkening of one side of terrain features, such as hills and ridges. The darker the shading, the steeper the slope. Shaded relief is sometimes used in conjunction with contour lines to emphasize these features.

d. **Hachures.** Hachures are short, broken lines used to show relief. Hachures are sometimes used with contour lines. They do not represent exact elevations, but are mainly used to show large, rocky outcrop areas. Hachures are used extensively on small-scale maps to show mountain ranges, plateaus, and mountain peaks.

e. **Contour Lines.** Contour lines are the most common method of showing relief and elevation on a standard topographic map. A contour line represents an imaginary line on the ground, above or below sea level. All points on the contour line are at the same elevation. The elevation represented by contour lines is the vertical distance above or below sea level. The three types of contour lines (Figure 10-1) used on a standard topographic map are as follows:

(1) *Index.* Starting at zero elevation or mean sea level, every fifth contour line is a heavier line. These are known as index contour lines. Normally, each index contour line is numbered at some point. This number is the elevation of that line.

(2) *Intermediate.* The contour lines falling between the index contour lines are called intermediate contour lines. These lines are finer and do not have their elevations given. There are normally four intermediate contour lines between index contour lines.

(3) *Supplementary.* These contour lines resemble dashes. They show changes in elevation of at least one half the contour interval. These lines are normally found where there is very little change in elevation, such as on fairly level terrain.

Figure 10-1. Contour lines

10-3. CONTOUR INTERVALS

Before the elevation of any point on the map can be determined, the user must know the contour interval for the map he is using. The contour interval measurement given in the marginal information is the vertical distance between adjacent contour lines. To determine the elevation of a point on the map—

a. Determine the contour interval and the unit of measure used, for example, feet, meters, or yards (Figure 10-2).

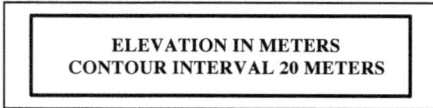

```
ELEVATION IN METERS
CONTOUR INTERVAL 20 METERS
```

Figure 10-2. Contour interval note.

b. Find the numbered index contour line nearest the point of which you are trying to determine the elevation (Figure 10-3).

c. Determine if you are going from lower elevation to higher, or vice versa. In Figure 10-3, point (a) is between the index contour lines. The lower index contour line is numbered 500, which means any point on that line is at an elevation of 500 meters above mean sea level. The upper index contour line is numbered 600, or 600 meters. Going from the lower to the upper index contour line shows an increase in elevation.

d. To determine the exact elevation of point (a), start at the index contour line numbered 500 and count the number of intermediate contour lines to point (a). Point (a) is located on the second intermediate contour line above the 500-meter index contour line. The contour interval is 20 meters (Figure 10-2), thus each one of the intermediate contour lines crossed to get to point (a) adds 20 meters to the 500-meter index contour line. The elevation of point (a) is 540 meters; the elevation has increased.

e. To determine the elevation of point (b), go to the nearest index contour line. In this case, it is the upper index contour line numbered 600. Point (b) is located on the intermediate contour line immediately below the 600-meter index contour line. Below means downhill or a lower elevation. Therefore, point (b) is located at an elevation of 580 meters. Remember, if you are decreasing elevation, add the contour interval to the nearest index contour line. If you are decreasing elevation, subtract the contour interval from the nearest index contour line.

f. To determine the elevation to a hilltop, point (c), add one-half the contour interval to the elevation of the last contour line. In this example, the last contour line before the hilltop is an index contour line numbered 600. Add one-half the contour interval, 10 meters, to the

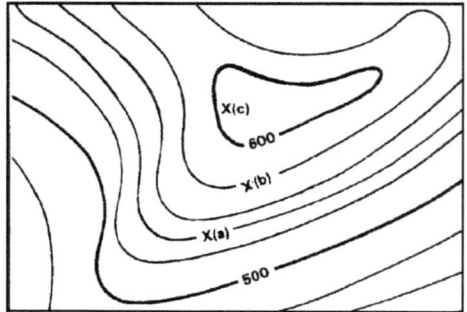

Figure 10-3. Points on contour lines

index contour line. The elevation of the hilltop would be 610 meters.

g. There may be times when you need to determine the elevation of points to a greater accuracy. To do this, you must determine how far between the two contour lines the point lies. However, most military needs are satisfied by estimating the elevation of points between contour lines (Figure 10-4).

(1) If the point is less than one-fourth the distance between contour lines, the elevation will be the same as the last contour line. In Figure 10-4, the elevation of point (a) will be 100 meters. To estimate the elevation of a point between one-fourth and three-fourths of the distance between contour lines, add one-half the contour interval to the last contour line.

(2) Point (b) is one-half the distance between contour lines. The contour line immediately below point (b) is at an elevation of 160 meters. The contour interval is 20 meters; thus one-half the contour interval is 10 meters. In this case, add 10 meters to the last contour line of 160 meters. The elevation of point (b) would be approximately 170 meters.

Figure 10-4. Points between contour lines.

(3) A point located more than three-fourths of the distance between contour lines is considered to be at the same elevation as the next contour line. Point (c) is located three-fourths of the distance between contour lines. In Figure 10-4, point (c) would be considered to be at an elevation of 180 meters.

h. To estimate the elevation to the bottom of a depression, subtract one-half the contour interval from the value of the lowest contour line before the depression. In Figure 10-5, the lowest contour line before the depression is 240 meters in elevation. Thus, the elevation at the edge of the depression is 240 meters. To determine the elevation at the bottom of the depression, subtract one-half the contour interval. The contour interval for this example is 20 meters. Subtract 10 meters from the lowest contour line immediately before the depression. The result is that the elevation at the bottom of the depression is 230 meters. The tick marks on the contour line forming a depression always point to lower elevations.

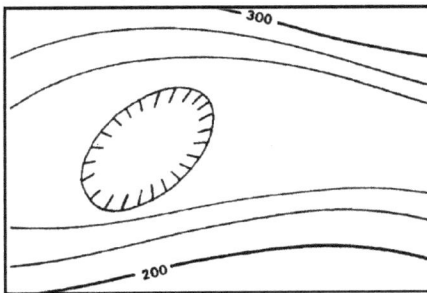

Figure 10-5. Depression.

i. In addition to the contour lines, bench marks and spot elevations are used to indicate points of known elevations on the map.

(1) Bench marks, the more accurate of the two, are symbolized by a black X, such as X BM 214. The 214 indicates that the center of the X is at an elevation of 214 units of measure (feet, meters, or yards) above mean sea level. To determine the units of measure, refer to the contour interval in the marginal information.

(2) Spot elevations are shown by a brown X and are usually located at road junctions and on hilltops and other prominent terrain features. If the elevation is shown in black numerals, it has been checked for accuracy; if it is in brown, it has not been checked.

NOTE: New maps are being printed using a dot instead of brown Xs.

10-4. TYPES OF SLOPES

Depending on the military mission, soldiers may need to determine not only the height of a hill, but the degree of the hill's slope as well. The rate of rise or fall of a terrain feature is known as its slope. The speed at which equipment or personnel can move is affected by the slope of the ground or terrain feature. This slope can be determined from the map by studying the contour lines—the closer the contour lines, the steeper the slope; the farther apart the contour lines, the gentler the slope. Four types of slopes that concern the military are as follows:

a. **Gentle**. Contour lines showing a uniform, gentle slope will be evenly spaced and wide apart (Figure 10-6). Considering relief only, a uniform, gentle slope allows the defender to use grazing fire. The attacking force has to climb a slight incline.

b. **Steep**. Contour lines showing a uniform, steep slope on a map will be evenly spaced, but close together. Remember, the closer the contour lines, the steeper the slope (Figure 10-7). Considering relief only, a uniform, steep slope allows the defender to use grazing fire, and the attacking force has to negotiate a steep incline.

Figure 10-6. Uniform, gentle slope.

Figure 10-7. Uniform, steep slope.

c. **Concave.** Contour lines showing a concave slope on a map will be closely spaced at the top of the terrain feature and widely spaced at the bottom (Figure 10-8). Considering relief only, the defender at the top of the slope can observe the entire slope and the terrain at the bottom, but he cannot use grazing fire. The attacker would have no cover from the defender's observation of fire, and his climb would become more difficult as he got farther up the slope.

d. **Convex.** Contour lines showing a convex slope on a map will be widely spaced at the top and closely spaced at the bottom (Figure 10-9). Considering relief only, the defender at the top of the convex slope can obtain a small distance of grazing fire, but he cannot observe most of the slope or the terrain at the bottom. The attacker will have concealment on most of the slope and an easier climb as he nears the top.

Figure 10-8. Concave slope.

Figure 10-9. Convex slope.

10-5. PERCENTAGE OF SLOPE

The speed at which personnel and equipment can move up or down a hill is affected by the slope of the ground and the limitations of the equipment. Because of this, a more exact way of describing a slope is necessary.

a. Slope may be expressed in several ways, but all depend upon the comparison of vertical distance (VD) to horizontal distance (HD) (Figure 10-10). Before we can determine the percentage of a slope, we must know the VD of the slope. The VD is determined by subtracting the lowest point of the slope from the highest point. Use the contour lines to determine the highest and lowest point of the slope (Figure 10-11).

b. To determine the percentage of the slope between points (a) and (b) in Figure 10-11, determine the elevation of point (b) (590 meters). Then determine the elevation of point (a) (380 meters). Determine the vertical distance between the two points by subtracting the elevation of point (a) from the elevation of point (b). The difference (210 meters) is the VD between points (a) and (b). Then measure the HD between the two points on the map in Figure 10-12. After the horizontal distance has been determined, compute the percentage of the slope by using the formula shown in Figure 10-13.

Figure 10-10. Slope diagram.

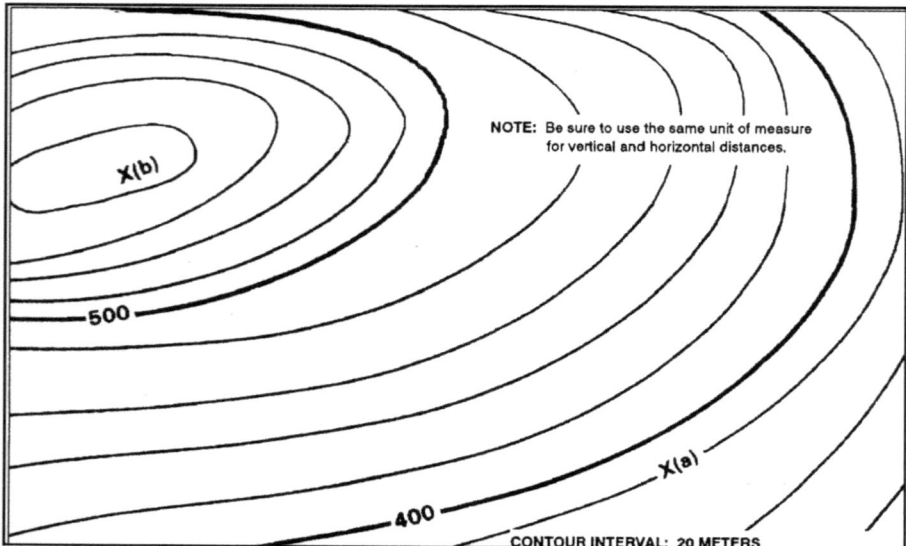

NOTE: Be sure to use the same unit of measure for vertical and horizontal distances.

CONTOUR INTERVAL: 20 METERS

Figure 10-11. Contour line around slope.

Figure 10-12. Measuring horizontal distance

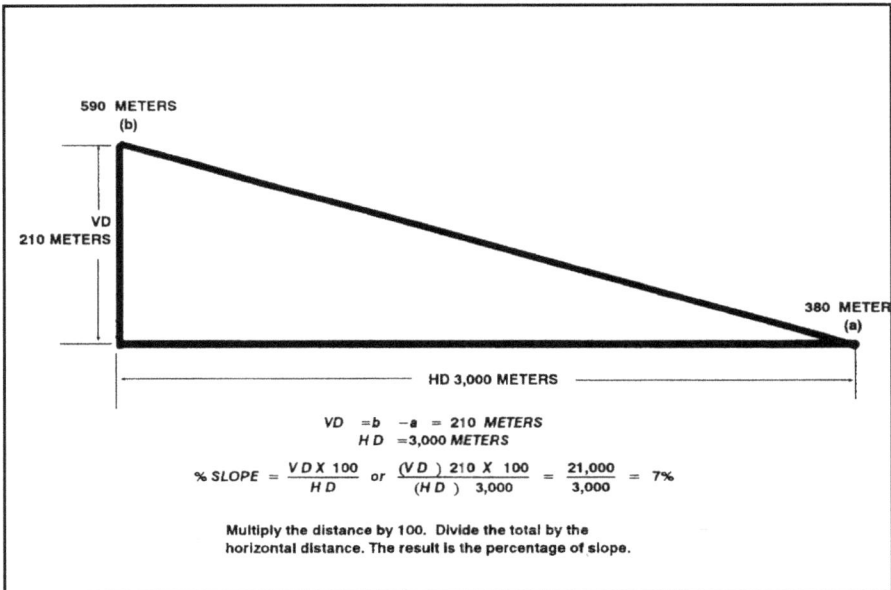

$$VD = b - a = 210 \ METERS$$
$$HD = 3,000 \ METERS$$

$$\% \ SLOPE = \frac{VD \times 100}{HD} \ or \ \frac{(VD)}{(HD)} \frac{210 \times 100}{3,000} = \frac{21,000}{3,000} = 7\%$$

Multiply the distance by 100. Divide the total by the
horizontal distance. The result is the percentage of slope.

Figure 10-13. Percentage of slope in meters.

c. The slope angle can also be expressed in degrees. To do this, determine the VD and HD of the slope. Multiply the VD by 57.3 and then divide the total by the HD (Figure 10-14). This method determines the approximate degree of slope and is reasonably accurate for slope angles less than 20°.

d. The slope angle can also be expressed as a gradient. The relationship of horizontal and vertical distance is expressed as a fraction with a numerator of one (Figure 10-15).

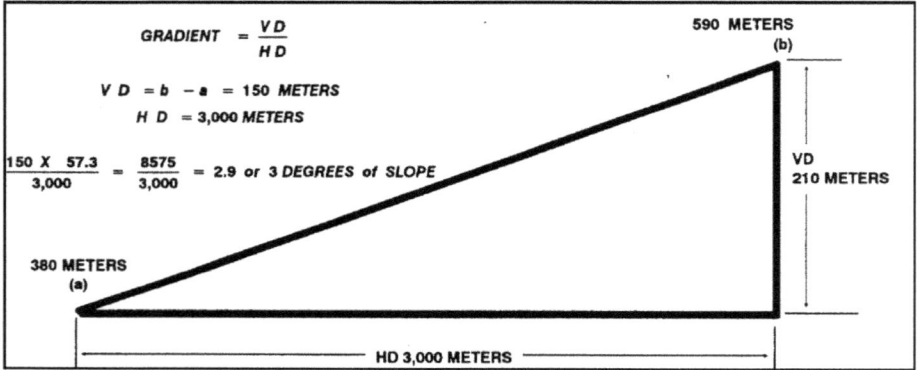

$$GRADIENT = \frac{VD}{HD}$$

$VD = b - a = 150\ METERS$

$HD = 3,000\ METERS$

$$\frac{150 \times 57.3}{3,000} = \frac{8575}{3,000} = 2.9\ or\ 3\ DEGREES\ of\ SLOPE$$

590 METERS (b)

VD 210 METERS

380 METERS (a)

HD 3,000 METERS

Figure 10-14. Degree of slope.

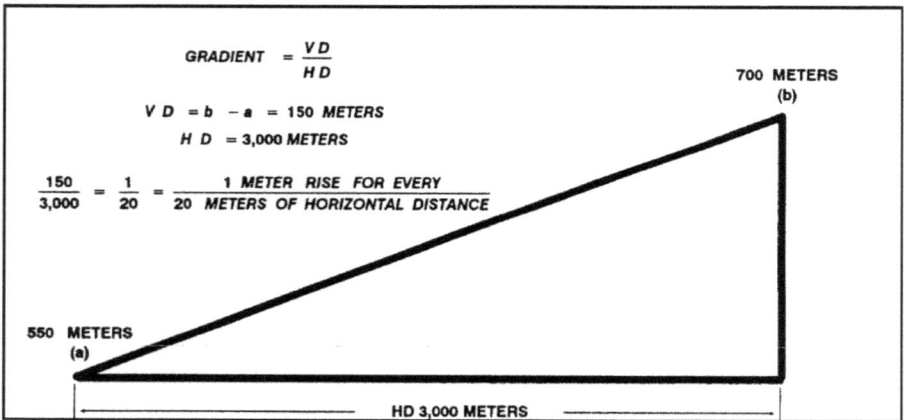

$$GRADIENT = \frac{VD}{HD}$$

$VD = b - a = 150\ METERS$

$HD = 3,000\ METERS$

$$\frac{150}{3,000} = \frac{1}{20} = \frac{1\ METER\ RISE\ FOR\ EVERY}{20\ METERS\ OF\ HORIZONTAL\ DISTANCE}$$

700 METERS (b)

550 METERS (a)

HD 3,000 METERS

Figure 10-15. Gradient.

10-6. TERRAIN FEATURES

All terrain features are derived from a complex landmass known as a mountain or ridgeline (Figure 10-16). The term ridgeline is not interchangeable with the term ridge. A ridgeline is a line of high ground, usually with changes in elevation along its top and low ground on all sides from which a total of 10 natural or manmade terrain features are classified.

Figure 10-16. Ridgeline.

a. **Major Terrain Features.**

(1) *Hill.* A hill is an area of high ground. From a hilltop, the ground slopes down in all directions. A hill is shown on a map by contour lines forming concentric circles. The inside of the smallest closed circle is the hilltop (Figure 10-17).

Figure 10-17. Hill.

(2) *Saddle.* A saddle is a dip or low point between two areas of higher ground. A saddle is not necessarily the lower ground between two hilltops; it may be simply a dip or break along a level ridgecrest. If you are in a saddle, there is high ground in two opposite directions and lower ground in the other two directions. A saddle is normally represented as an hourglass (Figure 10-18).

Figure 10-18. Saddle

(3) *Valley.* A valley is a stretched-out groove in the land, usually formed by streams or rivers. A valley begins with high ground on three sides, and usually has a course of running water through it. If standing in a valley, there is high ground in two opposite directions and a gradual inclination in the other two directions. Depending on its size and where a person is standing, it may not be obvious that there is high ground in the third direction, but water flows from higher to lower ground. Contour lines forming a valley are either U-shaped or V-shaped. To determine the direction water is flowing, look at the contour lines. The closed end of the contour line (U or V) always points upstream or toward high ground (Figure 10-19).

Figure 10-19. Valley

(4) *Ridge*. A ridge is a sloping line of high ground. If you are standing on the centerline of a ridge, you will normally have low ground in three directions and high ground in one direction with varying degrees of slope. If you cross a ridge at right angles, you will climb steeply to the crest and then descend steeply to the base. When you move along the path of the ridge, depending on the geographic location, there may be either an almost unnoticeable slope or a very obvious incline. Contour lines forming a ridge tend to be U-shaped or V-shaped. The closed end of the contour line points away from high ground (Figure 10-20).

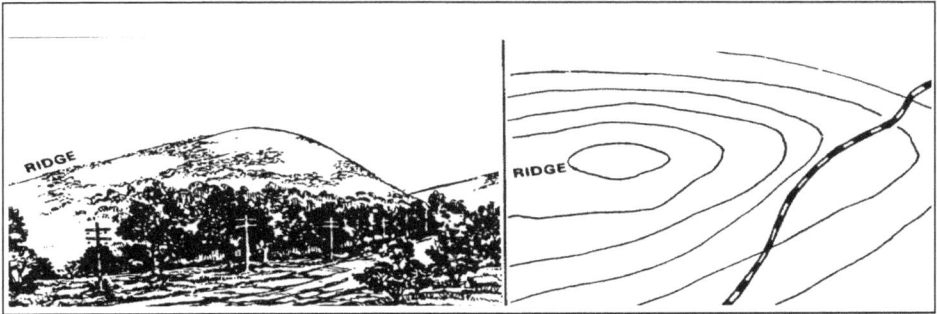

Figure 10-20. Ridge

(5) *Depression*. A depression is a low point in the ground or a sinkhole. It could be described as an area of low ground surrounded by higher ground in all directions, or simply a hole in the ground. Usually only depressions that are equal to or greater than the contour interval will be shown. On maps, depressions are represented by closed contour lines that have tick marks pointing toward low ground (Figure 10-21).

Figure 10-21. Depression

b. Minor Terrain Features

(1) *Draw.* A draw is a less developed stream course than a valley. In a draw, there is essentially no level ground and, therefore, little or no maneuver room within its confines. If you are standing in a draw, the ground slopes upward in three directions and downward in the other direction. A draw could be considered as the initial formation of a valley. The contour lines depicting a draw are U-shaped or V-shaped, pointing toward high ground (Figure 10-22).

Figure 10-22. Draw

(2) *Spur.* A spur is a short, continuous sloping line of higher ground, normally jutting out from the side of a ridge. A spur is often formed by two roughly parallel streams cutting draws down the side of a ridge. The ground will slope down in three directions and up in one. Contour lines on a map depict a spur with the U or V pointing away from high ground (Figure 10-23).

Figure 10-23. Spur.

(3) *Cliff*. A cliff is a vertical or near vertical feature, it is an abrupt change of the land. When a slope is so steep that the contour lines converge into one "carrying" contour of contours, this last contour line has tick marks pointing toward low ground (Figure 10-24A). Cliffs are also shown by contour lines very close together and, in some instances, touching each other (Figure 10-24B).

Figure 10-24A. Cliff.

Figure 10-24B. Cliff.

c. **Supplementary Terrain Features.**

(1) *Cut.* A cut is a man-made feature resulting from cutting through raised ground, usually to form a level bed for a road or railroad track. Cuts are shown on a map when they are at least 10 feet high, and they are drawn with a contour line along the cut line. This contour line extends the length of the cut and has tick marks that extend from the cut line to the roadbed, if the map scale permits this level of detail (Figure 10-25).

(2) *Fill.* A fill is a man-made feature resulting from filling a low area, usually to form a level bed for a road or railroad track. Fills are shown on a map when they are at least 10 feet high, and they are drawn with a contour line along the fill line. This contour line extends the length of the filled area and has tick marks that point toward lower ground. If the map scale permits, the length of the fill tick marks are drawn to scale and extend from the base line of the fill symbol (Figure 10-25).

Figure 10-25. Cut and fill.

10-7. INTERPRETATION OF TERRAIN FEATURES
Terrain features do not normally stand alone. To better understand these when they are depicted on a map, you need to interpret them. You can interpret terrain features (Figure 10-26) by using contour lines, the SOSES approach, ridgelining, or streamlining.

a. Contour Lines. Emphasizing the main contour lines is a technique used to interpret the terrain of an area. By studying these contour lines, you will get a better understanding of the layout of the terrain and be able to decide on the best route.

(1) The following description pertains to Figure 10-27. Running east to west across the complex landmass is a ridgeline. A ridgeline is a line of high ground, usually with changes in elevation along its top and low ground on all sides. The changes in elevation are the three hilltops and two saddles along the ridgeline. From the top of each hill, there is lower ground in all directions.

The saddles have lower ground in two directions and high ground in the opposite two directions. The contour lines of each saddle form half an hourglass shape. Because of the difference in size of the higher ground on the two opposite sides of a saddle, a full hourglass shape of a saddle may not be apparent.

(2) There are four prominent ridges. A ridge is on each end of the ridgeline and two ridges extend south from the ridgeline. All of the ridges have lower ground in three directions and higher ground in one direction. The closed ends of the U's formed by the contour lines point away from higher ground.

(3) To the south lies a valley; the valley slopes downward from east to west. Note that the U of the contour line points to the east, indicating higher ground in that direction and lower ground to the west. Another look at the valley shows high ground to the north and south of the valley.

(4) Just east of the valley is a depression. Looking from the bottom of the depression, there is higher ground in all directions.

(5) There are several spurs extending generally south from the ridgeline. They, like ridges, have lower ground in three directions and higher ground in one direction. Their contour line U's point away from higher ground.

(6) Between the ridges and spurs are draws. They, like valleys, have higher ground in three directions and lower ground in one direction. Their contour line U's and V's point toward higher ground.

(7) Two contour lines on the north side of the center hill are touching or almost touching. They have ticks indicating a vertical or nearly vertical slope or a cliff.

(8) The road cutting through the eastern ridge depicts cuts and fills. The breaks in the contour lines indicate cuts, and the ticks pointing away from the road bed on each side of the road indicate fills.

b. **SOSES**. A recommended technique for identifying specific terrain features and then locating them on the map is to make use of five of their characteristics known by the mnemonic SOSES. Terrain features can be examined, described, and compared with each other and with corresponding map contour patterns in terms of their shapes, orientations, sizes, elevations, and slopes.

(1) *Shape*. The general form or outline of the feature at its base.

(2) *Orientation*. The general trend or direction of a feature from your viewpoint. A feature can be in line, across, or at an angle to your viewpoint.

(3) *Size*. The length or width of a feature horizontally across its base. For example, one terrain feature might be larger or smaller than another.

(4) *Elevation*. The height of a terrain feature. This can be described either in absolute or relative terms as compared to the other features in the area. One landform may be higher, lower, deeper, or shallower than another.

(5) *Slope*. The type (uniform, convex, or concave) and the steepness or angle (steep or gentle) of the sides of a terrain feature.

Through practice, you can learn to identify several individual terrain features in the field and see how they vary in appearance.

| 1. HILL | 3. RIDGE | 5. DEPRESSION | 7. SPUR | 9. CUT |
| 2. VALLEY | 4. SADDLE | 6. DRAW | 8. CLIFF | 10. FILL |

Figure 10-26. Terrain features.

NOTE: Further terrain analysis using SOSES can be learned by using the Map Interpretation and Terrain Association Course. It consists of three separate courses of instruction: basic, intermediate, and advanced. Using photographic slides of terrain and other features, basic instruction teaches how to identify basic terrain feature types on the ground and on the map. Intermediate instruction teaches elementary map interpretation and terrain association using real world scenes and map sections of the same terrain. Advanced instruction teaches advanced techniques for map interpretation and terrain association. The primary emphasis is on the concepts of map design guidelines and terrain association skills. Map design guidelines refer to the rules and practices used by cartographers in the compilation and symbolization of military topographic maps. Knowledge of the selection, classification, and symbolization of mapped features greatly enhances the user's ability to interpret map information.

c. **Ridgelining.** This technique helps you to visualize the overall lay of the ground within the area of interest on the map. Follow these steps:

(1) Identify on the map the crests of the ridgelines in your area of operation by identifying the close-out contours that lie along the hilltop.

(2) Trace over the crests so each ridgeline stands out

clearly as one identifiable line.

(3) Go back over each of the major ridgelines and trace over the prominent ridges and spurs that come out of the ridgelines. The usual colors used for this tracing are red or brown; however, you may use any color at hand. When you have completed the ridgelining process, you will find that the high ground on the map will stand out and that you will be able to see the relationship between the various ridge- lines (Figure 10-27).

d. **Streamlining.** This procedure (Figure 10-27) is similar to that of ridgelining.

(1) Identify all the mapped streams in the area of operations.

(2) Trace over them to make them stand out more prominently.

(3) Then identify other low ground, such as smaller valleys or draws that feed into the major streams, and trace over them. This brings out the drainage pattern and low ground in the area of operation on the map. The color used for this is usually blue; but again, if blue is not available, use any color at hand so long as the distinction between the ridgelines and the streamlines is clear.

10-8. PROFILES

The study of contour lines to determine high and low points of elevation is usually adequate for military op-

Figure 10-27. Ridgelining and streamlining.

erations. However, there may be a few times when we need a quick and precise reference to determine exact elevations of specific points. When exactness is demanded, a profile is required. A profile, within the scope and purpose of this manual, is an exaggerated side view of a portion of the earth's surface along a line between two or more points.

a. A profile can be used for many purposes. The primary purpose is to determine if line of sight is available. Line of sight is used—

(1) To determine defilade positions.

(2) To plot hidden areas or dead space.

(3) To determine potential direct fire weapon positions.

(4) To determine potential locations for defensive positions.

(5) To conduct preliminary planning in locating roads, pipelines, railroads, or other construction projects.

b. A profile can be constructed from any contoured map. Its construction requires the following steps:

(1) Draw a line on the map from where the profile is to begin to where it is to end (Figure 10-28).

(2) Find the value of the highest and lowest contour lines that cross or touch the profile line. Add one contour value above the highest and one below the lowest to take care of hills and valleys.

(3) Select a piece of lined notebook paper with as many lines as was determined in (2) above. The standard Army green pocket notebook or any other paper with 1/4-inch lines is ideal. Wider lines, up to 5/8-inch, may be used. If lined paper is not available draw equally spaced horizontal lines on a blank sheet of paper.

(4) Number the top line with the highest value and the bottom line with the lowest value as determined in (2) above.

Figure 10-28. Connecting points.

(5) Number the rest of the lines in sequence, starting with the second line from the top. The lines will be numbered in accordance with the contour interval (Figure 10-29).

(6) Place the paper on the map with the lines next to and parallel to the profile line (Figure 10-29).

(7) From every point on the profile line where a contour line, stream, intermittent stream, or other body of water crosses or touches, drop a perpendicular line to the line having the same value. Place a tick mark where the perpendicular line crosses the number line (Figure 10-29). Where trees are present, add the height of the trees to the contour line and place a tick mark there. Assume the height of the trees to be 50 feet or 15 meters where dark green tint is shown on the map. Vegetation height may be adjusted up or down when operations in the area have provided known tree heights.

Figure 10-29. Dropping perpendiculars.

(8) After all perpendicular lines have been drawn and tick marks placed where the lines cross, connect all tick marks with a smooth, natural curve to form a horizontal view or profile of the terrain along the profile line (Figure 10-29).

(9) The profile drawn may be exaggerated. The spacing between the lines drawn on the sheet of paper will determine the amount of exaggeration and may be varied to suit any purpose.

(10) Draw a straight line from the start point to the end point on the profile. If the straight line intersects the curved profile, line of sight to the end point is not available (Figure 10-30).

(11) Line of sight to other points along the profile line can be determined by drawing a line from the start point to additional points. In Figure 10-30, line of sight is available to—

A-Yes	**D-Yes**	**G-Yes**
B-No	**E-No**	**H-No**
C-No	**F-No**	**I-No**

(12) The vertical distance between navigable ground up to the line of sight line is the depth of defilade.

Figure 10-30. Drawing lines to additional points.

c. When time is short, or when a complete profile is not needed, one may be constructed showing only the hilltops, ridges, and if desired, the valleys. This is called a hasty profile. It is constructed in the same manner as a full profile (Figure 10-31).

Figure 10-31. Drawing a hasty profile.

TERRAIN ASSOCIATION

Failure to make use of the vast amounts of information presented by the map and available to the eye on the ground will seriously reduce your chances for success in land navigation. the soldier who has repeatedly practiced the skills of identifying and discriminating among the many types of terrain an other features, knows how these features are mapped, can begin to value the shape of the land by studying the map, can estimate distances, and can perform quick resection from the many landmarks he sees is the one who will be at the right place to help defeat the enemy on the battlefield. This chapter tells how to orient a map with and without a compass, how to find locations on a map as well as on the ground, how to study the terrain, and how to move on the ground using terrain association and dead reckoning.

11-1. ORIENTING THE MAP

The first step for a navigator in the field is orienting the map. A map is oriented when it is in a horizontal position with its north and south corresponding to the north and south on the ground. Some orienting techniques follow:

a. **Using a Compass**. When orienting a map with a compass, remember that the compass measures magnetic azimuths. Since the magnetic arrow points to magnetic north, pay special attention to the declination diagram. There are two techniques.

(1) *First technique*. Determine the direction of the declination and its value from the declination diagram.

(a) With the map in a horizontal position, take the straightedge on the left side of the compass and place it alongside the north-south gad line with the cover of the compass pointing toward the top of the map. This will place the fixed black index line of the compass parallel to north-south and lines of the map.

(b) Keeping the compass aligned as directed above, rotate the map and compass together until the magnetic arrow is belong the fixed black index line on the compass. At this time, the map is close to being oriented.

(c) Rotate the map and compass in the direction of the declination diagram.

(d) If the magnetic north arrow on the map is to the left of the grid north, the compass reading will equal the G-M angle given in the declination diagram. The map is then oriented (Figure 11-1).

Figure 11-1. Map oriented with 11 degrees west declination

Figure 11-2. Map oriented with 21 degrees east declination.

Figure 11-3. Map oriented with 15 degrees east declination.

(e) If the magnetic north is to the right of grid north, the compass reading wilt equal 360° minus the G-M angle (Figure 11-2).

(2) Second technique. Determine the direction of the declination and its value from the declination diagram.

(a) Using any north-south grid line on the map as a base, draw a magnetic azimuth equal to the G-M angle given in the declination diagram with the protractor.

(b) If the declination is easterly (right), the drawn line is equal to the value of the G-M angle. Then align the straightedge, which is on the left side of the compass, alongside the drawn line on the map. Rotate the map and compass until the magnetic arrow of the compass is below the fixed black index line. The map is now oriented (Figure 11-3).

(c) If the declination is westerly (left), the drawn line will equal 360° minus the value of the G-M angle. Then align the straightedge, which is on the left side of the compass, alongside the drawn line on the map. Rotate the map and compass until the magnetic arrow of the com- pass is below the fixed black index line. The map is now oriented (Figure 11-4).

NOTE: Once the map is oriented, magnetic azimuths can be determined with the compass, but the map should not be moved from its oriented position; any change in its position will move it out of line with magnetic north. (See paragraph 11-6b[1]).

NOTE: Special care should be taken whenever orienting your map with a compass. A small mistake can cause you to navigate in the wrong direction.

b. Using Terrain Association. A map can be oriented by terrain association when a compass is not available or when the user has to make many quick references as he moves across country. Using this method requires careful examination of the map and the ground, and the user must know his approximate location (Figure 11-5). Orienting by this method is discussed in detail in paragraph 11-3.

Figure 11-4 Map oriented with 10 degrees west declination..

Figure 11-5. Terrain association.

c. **Using Field-Expedient Methods.**
When a compass is not available and there
are no recognizable terrain features, a map
may be oriented by any of the field-expedient
methods described in paragraph 9-5. Also see
Figure 11-6

Figure 11-6. Field-expedient method

11-2. LOCATIONS

The key to success in land navigation is to know your
location at all times. With this basic knowledge,you can
decide what direction and what distance to travel.

a. **Know Position.** Most important of all is the initial
location of theuser before starting any movement in the field.
If movement takes place without establishing the initial
location, everthing that is done in the field from there on is a
gamble. Determine the initial location by referring to the last
known position, by grid coordinates and terrain association,
or by locating and orienting your position on the map and
the ground.

b. **Known Point/Known Distance (Polar Plot).** This
location can be determined by knowing the starting point,
the azimuth to the desired objective, and the distance to it.

c. **Resection.** See chapter 6.

d. **Modified Resection** See Chapter 6.

e. **Intersection.** See chapter 6.

f. **Indirect Fire.** Finding a location by indirect
fore is done with smoke. Use the point of impacts of the
rounds as a reference point from which distances amd
azimuth can be obtained.

11-3. TERRAUB ASSOCIATION USAGE

This technique of moving by terrain association is more
forgiving of mistakes and far less time-consuming than
dead reckoning. It best suits those situations that call for
movement from one area to another. Errors made using
terrain association are easily corrected because you are
comparing what you expected to see from the map to
what you do see on the ground. Erros are anticipated and
will not go unchecked. You can easily make adjustment.

based upon what you encounter. Periodic position-fixing through either plotted or estimated resection will also make it possible to correct your movements, call for fire, or call in the locations of enemy targets or any other information of tactical or logistical importance.

a. **Matching the Terrain to the Map by Examining Terrain Features.** By observing the contour lines in detail, the five mayor terrain features (hilltop, valley, ridge, depression, and saddle) should be determined. This is a simple task in an area where the observer has ample view of the terrain in all directions. One by one, match the terrain features depicted on the map with the same features on the ground. In restricted terrain, this procedure becomes harder; however, constant checking of the map as you move is the determining factor (Figure 11-5, page 11-3).

b. **Comparing the Vegetation Depicted on the Map.** When comparing the vegetation, a topographic map should be used to make a comparison of the clearings that appear on the map with the ones on the ground. The user must be familiar with the different symbols, such as vineyards, plantations, and orchards that appear on the legend. The age of the map is an important factor when comparing vegetation. Some important vegetation features were likely to be different when the map was made. Another important factor about vegetation is that it can change overnight by natural accidents or by man (forest fires, clearing of land for new developments, farming, and so forth).

c. **Masking by the Vegetation.** Important landforms could be camouflaged by the vegetation, making it harder for the navigator to use terrain association.

d. **Using the Hydrography.** Inland bodies of water can help during terrain association. The shape and size of lakes in conjunction with the size and direction of flow of the rivers and streams are valuable help.

e. **Using man-made Features.** Man-made features could be an important factor during terrain association. The user must be familiar with the symbols shown in the legend representing those features. The direction of buildings, roads, bridges, high-tension lines, and so forth will make the terrain inspection a lot easier; however, the age of the map must be considered because man-made features appear and disappear constantly.

f. **Examining the Same Piece of Terrain During the Different Seasons of the Year.** In those areas of the world where the seasons are very distinctive, a detailed examination of the terrain should be made during each of the seasons. The same piece of land will not present the same characteristics during both spring and winter.

(1) During winter, the snow will pack the vegetation, delineating the land, making the terrain features appear as clear as they are shown by the

contour lines on the map. Ridges, valleys, and saddles are very distinctive.

(2) During spring, the vegetation begins to reappear and grow. This causes a gradual change of the land to the point that the foliage conceals the terrain features and makes the terrain hard to recognize.

(3) During summer months, the effects are similar to those in the spring.

(4) Fall will make the land appear different with its change of color and gradual loss of vegetation.

(5) During the rainy season, the vegetation is green and thick, and the streams and ponds will look like small rivers and lakes. In scarcely vegetated areas, the erosion will change the shape of the land.

(6) During a period of drought, the vegetation dries out and becomes vulnerable to forest fires that change the terrain whenever they occur. Also, during this season the water levels of streams and lakes drop, adding new dimensions and shape to the existing mapped areas.

g. **Following an Example of Terrain Association.** Your location is hilltop 514 in the lower center of the map in Figure 11-7.

(1) *To the north.* The contour lines indicate that the hill slopes down for about 190 meters, and that it leads into a small valley containing an intermittent stream. On the other side of the stream as you continue with your northerly inspection, the terrain will start a gradual ascent, indicating a hilltop partially covered with vegetation, until an unimproved road is reached. This road runs along a gradual ridgeline with north-west direction. Then the contour lines spacing narrow, indicating a steeper grade that leads to a narrow valley containing a small intermittent stream. As you continue up, you find a small but prominent ridge with a clearing. The contour lines once again show a steeper grade leading to a moderate valley containing an intermittent stream running in a south-east direction.

(2) *To the east.* There is a clearing of the terrain as it slopes down to Schley Pond. An ample valley is clearly seen on the right side of the pond, as indicated by the "U" and "V" shape of the contour lines. This valley contains some swamp areas and there is a long ridgeline on the north portion of the valley.

(3) *To the south.* The terrain gently slopes downward until a clear area is reached. It continues in a downward direction to an intermittent stream running south-east in a small valley. There is also an improved road running in the same direction as the valley. At the intersection of the roads as you face south, there is a clearing of about 120 meters on the ridge. At the bottom of it, a stream runs from Schley Pond in a south-west direction through an ample valley fed by two intermittent streams. As you

Figure 11-7. Example of terrain association

continue, a steep, vegetated hill is found with a clearing on its top, followed by a small saddle an another hilltop.

(4) To the west. First, you see a small, clear valley. It is followed by a general ridgeline running north-west in which an unimproved road is located just before a hilltop. Continuing on a westerly direction, you will find a series of alternate valleys and ridges.

11-4. TACTICAL CONSIDERATIONS
Military cross-country navigation is intellectually demanding because it is imperative that the unit, crew, or vehicle survive and successfully complete the move in order to accomplish its mission. However, the unnecessary use of a difficult route will make navigation too complicated, create more noise when proceeding over it, cause wear and tear on equipment and personnel, increase the need for and needlessly complicate recovery operations, and waste scarce time. On receipt of a tactical mission, the leader begins his troop-leading

procedures and makes a tentative plan. He bases the tentative plan on a good terrain analysis. He analyzes the considerations covered in the following mnemonics—OCOKA and METT-T.

a. **OCOKA.** The terrain should be analyzed for observation and fields of fire, cover and concealment, obstacles, key terrain, and avenues of approach.

(1) *Observation and fields of fire.* The purpose of observation is to see the enemy (or various landmarks) but not be seen by him. Anything that can be seen can be hit. Therefore, a field of hire is an area that a weapon or a group of weapons can cover effectively with fire from a given position.

(2) *Cover and concealment* Cover is shelter or protection (from enemy fire) either natural or artificial. Always try to use covered routes and seek cover for each halt, no matter how brief it is planned to be. Unfortunately, two factors interfere with obtaining constant cover. One is time and the other is terrain. Concealment

is protection from observation or surveillance, including concealment from enemy air observation. Before, trees provided good concealment, but with modern thermal and infrared imaging equipment, trees are not always effective. When you are moving, concealment is generally secondary; select routes and positions that will not allow a covered or concealed enemy near you.

(3) *Obstacles*. These are any obstructions that stop, delay, or divert movement. Obstacles can be natural (rivers, swamps, cliffs, or mountains) or they may be artificial (barbed wire entanglements, pits, concrete or metal antimechanized traps). They may be ready-made or they may be constructed in the field. Always consider any possible obstacles along your movement route and, if possible, try to keep obstacles between the enemy and yourself.

(4) *Key terrain*. This is any locality or area that the seizure or retention of affords a marked advantage to either combatant. Urban areas are often seen by higher headquarters as being key terrain because they can be used to control routes. On the other hand, an urban area that is destroyed may be an obstacle instead. High ground can be key because it dominates an area with good observation and fields of fire. In an open area, a draw or wadi (dry streambed located in an arid area) may provide the only cover for many kilometers, thereby becoming key. You should always attempt to locate any area near you that could be even remotely considered as key terrain.

(5) *Avenues of approach*. These are access routes. They may be the routes you can use to get to the enemy or the routes they can use to get to you. Basically, an identifiable route that approaches a position or location is an avenue of approach to that location. They are often terrain corridors such as valleys or wide, open areas.

b. METT-T. Tactical factors other than the military aspects of terrain must also be considered in conjunction with terrain during movement planning and execution as well. These additional considerations are mission, enemy, terrain and weather, troops, and time available.

(1) *Mission*. This refers to the specific task assigned to a unit or individual. It is the duty or task, together with the purpose, that clearly indicates the action to be taken and the reason for it—but not how to do it. Training exercises should stress the importance of a thorough map reconnaissance to evaluate the terrain. This allows the leader to confirm his tentative plan, basing his decision on the terrain's effect on his mission.

(a) Marches by foot or vehicle are used to move troops from one location to another. Soldiers must get to the right place, at the right time, and in good fighting condition. The normal rate for an 8 hour foot march is 4 kilometers per hour. However, the rate of march may vary, depending on the following factors:
- Distance.
- Time allowed.
- Likelihood of enemy contact.
- Terrain.
- Weather.

Physical condition of soldiers.

Equipment/weight to be carried.

A motor march requires little or no walking by the soldiers, but the factors affecting the rate of march still apply.

(b) Patrol missions are used to conduct combat or reconnaissance operations. Without detailed planning and a thorough map reconnaissance, any patrol mission may not succeed. During the map reconnaissance, the mission leader determines a primary and alternate route to and from the objective(s).

(c) Movement to contact is conducted whenever an element is moving toward the enemy but is not in contact with the enemy. The lead element must orient its movement on the objective by conducting a map reconnaissance, determining the location of the objective on both the map and the ground, and selecting the route to be taken.

(d) Delays and withdrawals are conducted to slow the enemy down without becoming decisively engaged, or to assume another mission. To be effective, the element leader must know where he is to move and the route to be taken.

(2) *Enemy*. This refers to the strength, status of training, disposition (locations), doctrine, capabilities, equipment (including night vision devices), and probable courses of action which impact upon both the planning and execution of the mission, including a movement.

(3) *Terrain and weather*. Observation and fields of fire influence the placement of positions and crew served weapons. The leader conducts a map reconnaissance to determine key terrain, obstacles, cover and concealment, and likely avenues of approach.

(a) Key terrain is any area whose control affords a marked advantage to the force holding it. Some types of key terrain are high ground, bridges, towns, and road junctions.

(b) Obstacles are natural or man-made terrain features that stop, slow down, or divert movement. Consideration of obstacles is influenced by the unit's mission. An obstacle may be an advantage or disadvantage, depending upon the direction of attack or defense. Obstacles can be found by conducting a thorough map reconnaissance and study of recent aerial photographs.

(c) Cover and concealment must be determined for both friendly and enemy forces. Concealment is protection from observation; cover is protection from the effects of fire. Most terrain features that offer cover also provide concealment from ground observation. There are areas that provide no concealment from enemy observation. These danger areas maybe large or small open fields, roads, or streams. During the leader's map reconnaissance, he should determine any obvious danger areas and, if possible, adjust his route.

(d) Avenues of approach are routes by which a unit may reach an objective or key terrain. To be considered an AA, a route must provide enough width for the deployment of the size force for which it is being considered. AAs are also considered for the subordinate enemy force. For example, a company would determine likely AAs for an enemy platoon; a platoon would determine likely AAs for an enemy squad. Likely AAs may be either ridges, valleys, or by air. By examining the terrain, the leader can determine the likely enemy AAs based on the tactical situation.

(e) Weather has little impact on dismounted land navigation. Rain and snow could possibly slow down the rate of march, that is all. But during mounted land navigation, the navigator must know the effect of weather on his vehicle. (See Chapter 12 for mounted land navigation.)

(4) *Troops.* Consideration of your own troops is equally important. The size and type of the unit to be moved and its capabilities, physical condition, status of training, and types of equipment assigned will all affect the selection of routes, positions, fire plans, and the various decisions to be made during movement. On ideal terrain such as relatively level ground with little or no woods, a platoon can defend a front of up to 400 meters. The leader must conduct a thorough map reconnaissance and terrain analysis of the area his unit is to defend. Heavily wooded areas or very hilly areas may reduce the front a platoon can defend. The size of the unit must also be taken into consideration when planning a movement to contact. During movement, the unit must retain its ability to maneuver. A small draw or stream may reduce the unit's maneuverability but provide excellent concealment. All of these factors must be considered.

(a) Types of equipment that may be needed by the unit can be determined by a map reconnaissance. For example, if the unit must cross a large stream during its movement to the objective, ropes may be needed for safety lines.

(b) Physical capabilities of the soldiers must be considered when selecting a route. Crossing a large swampy area may present no problem to a physically fit unit, but to a unit that has not been physically conditioned, the swampy area may slow or completely stop its movement.

(5) *Tune available.* At times, the unit may have little time to reach an objective or to move from one point to another. The leader must conduct a map reconnaissance to determine the quickest route to the objective; this is not always a straight route. From point A to point B on the map may appear to be 1,000 meters, but if the route is across a large ridge, the distance will be greater. Another route from point A to B may be 1,500 meters — but on flat terrain. In this case, the quickest route would be across the flat terrain; however, concealment and cover may be lost.

11-5. MOVEMENT AND ROUTE SELECTION

One key to success in tactical missions is the ability to move undetected to the objective. There are four steps to land navigation. Being given an objective and the requirement to move there, you must know where you are, plan the route, stay on the route, and recognize the objective.

a. **Know Where You Are (Step 1).** You must know where you are on the map and on the ground at all times and in every possible way. This includes knowing where you are relative to—

- Your directional orientation.
- The direction and distances to your objective.
- Other landmarks and features.
- Any impassable terrain, the enemy, and danger areas.
- Both the advantages and disadvantages presented by the terrain between you and your objective. This step is accomplished by knowing how to read a map, recognize and identify specific terrain and other features; determine and estimate direction; pace, measure, and estimate distances, and both plot and estimate a position by resection.

b. **Plan the Route (Step 2).** Depending upon the size of the unit and the length and type of movement to be conducted, several factors should be considered in selecting a good route or routes to be followed. These include—

- Travel time.
- Travel distance. Maneuver room needed.
- Trafficability.
- Load-bearing capacities of the soil.
- Energy expenditure by troops.
- The factors of METT-T.
- Tactical aspects of terrain (OCOKA).
- Ease of logistical support.
- Potential for surprising the enemy.
- Availability of control and coordination features.
- Availability of good checkpoints and steering marks.

In other words, the route must be the result of careful map study and should address the requirements of the mission, tactical situation, and time available. But it must also provide for ease of movement and navigation.

(1) Three route-selection criteria that are important for small-unit movements are cover, concealment, and the availability of reliable checkpoint features. The latter is weighted even more heavily when selecting the route for a night operation. The degree of visibility and ease of recognition (visual impact) are the key to the proper selection of these features.

(2) The best checkpoints are linear features that cross the route. Examples include perennial streams, hard-top roads, ridges, valleys, railroads, and power transmission lines. Next, it is best to select features that represent elevation changes of at least two contour intervals such as hills, depressions, spurs, and draws. Primary reliance upon cultural features and vegetation is cautioned against because they are most likely to have changed since the map was last revised.

(3) Checkpoints located at places where changes in direction are made mark your decision points. Be especially alert to see and recognize these features during movement. During preparation and planning, it is especially important to review the route and anticipate where mistakes are most likely to be made so they can be avoided.

(4) Following a valley floor or proceeding near (not on) the crest of a ridgeline will generally offer easy movement, good navigation checkpoints, and sufficient cover and concealment. It is best to follow terrain features whenever you can—not to fight them.

(5) A lost or a late arriving unit, or a tired unit that is tasked with an unnecessarily difficult move, will not contribute to the accomplishment of a mission. On the other hand, the unit that moves too quickly and carelessly into a destructive ambush or leaves itself open to air strikes will also have little impact. Careful planning and study are required each time a movement route is to be selected.

c. **Stay on the Route (Step** 3). In order to know that you are still on the correct route, you must be able to compare the evidence you encounter as you move according to the plan you developed on the map when you selected your route. This may include watching your compass reading (dead reckoning) or recognizing various checkpoints or landmarks from the map in their anticipated positions and sequences as you pass them (terrain association). Or, better still, it should be a combination of both.

d. **Recognize the Objective (Step** 4). The destination is rarely a highly recognizable feature such as a dominant hilltop or road junction. Such locations are seldom missed by even the most inexperienced navigators, but they are often dangerous places for soldiers to occupy. The relatively small, obscure places are most likely to be the destinations.

(1) Just how does one travel over unfamiliar terrain for moderate to great distances and know when he reaches the destination? One minor error, when many are possible, will cause the target to be missed.

(2) The answer is simple. Select a checkpoint (reasonably close to the destination) that is not so difficult to find or recognize. Then plan a short, fine-tuned last leg from the new expanded objective to the final destination. For example, you may be able to plan and execute the move as a series of sequenced movements from one checkpoint or landmark to another using both the terrain and a compass to keep you on the correct course. Finally, after arriving at the last checkpoint, you might follow a specific compass azimuth and pace off the relatively short, known distance to the final, pinpoint destination. This is called point navigation. A short movement out from a unit position to an observation post or to a coordination point may also be accomplished in the same manner.

11-6. NAVIGATION METHODS

Staying on the route is accomplished through the use of one or two navigation techniques—dead reckoning and terrain association. Each method will now be discussed in detail.

a. **Moving by Dead Reckoning.** Dead reckoning consists of two fundamental steps. The first is the use of a protractor and graphic scales to determine the direction and distance from one point to another on a map. The second step is the use of a compass and some means of measuring distance to apply this information on the ground. In other words, it begins with the determination of a polar coordinate on a map and ends with the act of finding it on the ground.

(1) Dead reckoning along a given route is the application of the same process used by a map maker as he establishes a measured line of reference upon which to construct the framework of his map. Therefore, triangulation exercises (either resection or intersection) can be easily undertaken by the navigator at any time to either determine or confirm precise locations along or near his route. Between these position-fixes, your location can be established by measuring or estimating the distance travelled along the azimuth being followed from the previous known point. You might use pacing, a vehicle odometer, or the application of elapsed time for this purpose, depending upon the situation.

(2) Most dead reckoned movements do not consist of single straight-line distances because you cannot ignore the tactical and navigational aspects of the terrain, enemy situation, natural and man-made obstacles, time,

and safety factors. Another reason most dead reckoning movements are not single straight-line distances is because compasses and pace-counts are imprecise measures. Error from them compounds over distance; therefore, you could soon be far afield from your intended route even if you performed the procedures correctly. The only way to counteract this phenomenon is to reconfirm your location by terrain association or resection. Routes planned for dead reckoning will generally consist of a series of straight-line distances between several check points with perhaps some travel running on or parallel to roads or trails.

(3) There are two advantages to dead reckoning. First, dead reckoning is easy to teach and to learn. Second, it can be a highly accurate way of moving from one point to another if done carefully over short distances, even where few external cues are present to guide the movements.

(4) Never walk with your compass open and held in front of you to move during daylight across open country along a specified magnetic azimuth. The compass will not remain steady or level, and it will not provide you with the most accurate readings when used this way. Instead, you must begin at the start point, face with your compass in the proper direction, sight on a landmark that is located on the correct azimuth to be followed, close your compass, proceed to that landmark, and repeat the process as many times as is necessary to complete that particular straight-line segment of the route.

(5) The landmarks selected for this purpose are called steering marks, and their selection is crucial to success in dead reckoning. Steering marks should never be determined from a map study. They are selected as the march progresses and are commonly on or near the highest points you can see along the azimuth line you are following when they are selected. They may be uniquely shaped trees, rocks, hilltops, posts, towers, buildings— anything that can be easily identified. If you do not see a good steering mark to the front, you might use a back azimuth to some feature behind you until a good steering mark appears out in front. Characteristics of a good steering mark are:

(a) It must have some characteristics about it, such as color, shade of color, size, or shape (preferably all four), that will assure you that it will continue to be recognized as you approach it.

(b) If several easily distinguished objects appear along your line of march, the best steering mark will be the most distant object. This will enable you to travel farther with fewer references to the compass. If you have many options, select the highest object. A higher mark is not as easily lost to sight as is a lower mark that blends into the background as you approach it. A steering mark should be continuously visible as you move toward it.

(c) Steering marks selected at night must have even more unique shapes than those selected during daylight. As darkness approaches, colors disappear and objects appear as black or gray silhouettes. Instead of seeing shapes, you begin to see only the general outlines that may appear to change as you move and see the objects from slightly different angles.

(6) Dead reckoning without natural steering marks is used when the area through which you are traveling is devoid of features, or when visibility is poor. At night, it may be necessary to send a member of the unit out in front of your position to create your own steering mark in order to proceed. His position should be as far out as possible to reduce the number of chances for error as you move. Arm-and-hand signals or a radio may be used in placing him on the correct azimuth. After he has been properly located, move forward to his position and repeat the process until some steering marks can be identified or until you reach your objective.

(7) When handling obstacles/detours on the route, follow these guidelines:

(a) Whenever an obstacle forces you to leave your original line of march and take up a parallel one, always return to the original line as soon as the terrain or situation will permit.

(b) To turn clockwise (right) 90°, you must add 90° to your original azimuth. To turn counterclockwise (left) 90° from your current direction, you must subtract 90° from your present azimuth.

(c) When making a detour, be certain that only paces taken toward the final destination are counted as part of your forward progress. They should not be confused with the local pacing that takes place perpendicular to the route in order to avoid the problem area and in returning to the original line of march after the obstacle has been passed.

(8) Sometimes a steering mark on your azimuth of travel can be seen across a swamp or some other obstacle to which you can simply walk out around. Dead reckoning can then begin at that point. If there is no obvious steering mark to be seen across the obstacle, perhaps one can be located to the rear. Compute a back azimuth to this point and later sight back to it once the obstacle has been passed in order to get back on track.

(9) You can use the deliberate offset technique. Highly accurate distance estimates and precision compass work may not be required if the destination or an intermediate checkpoint is located on or near a large linear feature that runs nearly perpendicular to your direction of travel. Examples include roads or highways, railroads, power transmission lines, ridges, or streams. In these cases, you should apply a deliberate error (offset) of about 10° to the azimuth you planned to follow and then move, using the lensatic compass as a guide, in

that direction until you encounter the linear feature. You will know exactly which way to turn (left or right) to find your destination or checkpoint, depending upon which way you planned your deliberate offset.

(10) Because no one can move along a given azimuth with absolute precision, it is better to plan a few extra steps than to begin an aimless search for the objective once you reach the linear feature. If you introduce your own mistake, you will certainly know how to correct it. This method will also cope with minor compass errors and the slight variations that always occur in the earth's magnetic field.

(11) There are disadvantages to dead reckoning. The farther you travel by dead reckoning without confirming your position in relation to the terrain and other features, the more error you will accumulate in your movements. Therefore, you should confirm and correct your estimated position whenever you encounter a known feature on the ground that is also on the map. Periodically, you should accomplish a resection triangulation using two or more known points to pinpoint and correct your position on the map. Pace counts or any type of distance measurement should begin anew each time your position is confirmed on the map.

(a) It is dangerous to select a single steering mark, such as a distant mountaintop, and then move blindly toward it. What will you do if you must suddenly call for fire support or a medical evacuation? You must periodically use resection and terrain association techniques to pinpoint your location along the way.

(b) Steering marks can be farther apart in open country, thereby making navigation more accurate. In areas of dense vegetation, however, where there is little relief, during darkness, or in fog, your steering marks must be close together. This, of course, introduces more chance for error.

(c) Finally, dead reckoning is time-consuming and demands constant attention to the compass. Errors accumulate easily and quickly. Every fold in the ground and detours as small as a single tree or boulder also complicate the measurement of distance.

b. **Moving by Terrain Association.** The technique of moving by terrain association is more forgiving of mistakes and far less time-consuming than dead reckoning. It best suits those situations that call for movement from one area to another (Figure 11-8). Once an error has been made in dead reckoning, you are off the track. Errors made using terrain association are easily corrected, however, because you are comparing what you expected to see from the map to what you do see on the ground. Errors are anticipated and will not go

Figure 11-8. Terrain association navigation.

unchecked. You can easily make adjustments based upon what you encounter. After all, you do not find the neighborhood grocery store by deed reckoning—you adjust your movements according to the familiar landmarks you encounter along the way (Figure 11-8). Periodic position-fixing through either plotted or estimated resection will also make it possible to correct your movements, call for fire, or call in the locations of enemy targets or any other information of tactical or logistical importance.

(1) *Identifying and locating selected features.* Being able to identify and locate selected features both on the map and on the ground is essential to your success in moving by terrain association. The following rules may prove helpful.

(a) Be certain the map is properly oriented as you move along your route and use the terrain and other features as guides. The orientation of the map must match the terrain or you will become completely confused.

(b) To locate and identify features being used to guide your movement, look for the steepness and shape of the slopes, the relative elevations of the various features, and the directional orientations in relation to your own position and to the position of the other features you can see.

(c) Make use of the additional cues provided by Hydrography, culture, and vegetation. All the information you can gather will assist you in making the move. The ultimate test and the best practice for this movement technique is to go out in the field and use it. The use of terrain, other natural features, and any man-made objects that appear both on the map and on the ground must be practiced at every opportunity. There is no other way to learn or retain this skill.

(2) *Using handrails, catching features and navigational attack points.* First, because it is difficult to dead reckon without error over long distances with your compass, the alert navigator can often gain assistance from the terrain.

(a) **Handrails** are linear features like roads or highways, railroads, power transmission lines, ridgelines, or streams that run roughly parallel to your direction of travel. Instead of using precision compass work, you can rough compass without the use of steering marks for as long as this feature travels with you on your right or left. It acts as a handrail to guide the way.

(b) Second, when you reach the point where either your route or the handrail changes direction, you must be aware that it is time to go your separate ways. Some prominent feature located near this point is selected to provide this warning. This is called a **catching feature;** it can also be used to tell you when you have gone to far.

(c) Third, the catching feature may also be your **navigational attack point;**this is the place where area navigation ends and point navigation begins. From this

last easily identified checkpoint, the navigator moves cautiously and precisely along a given azimuth for a specified distance to locate the final objective. The selection of this navigational attack point is important. A distance of 500 meters or less is most desirable.

Figure 11-8. Terrain association navigation.

(3) *Recognizing the disadvantages of terrain association.* The major disadvantage to navigation by terrain association is that you must be able to interpret the map and analyze the world around you. Recognition of terrain and other features, the ability to determine and estimate direction and distance, and knowing how to do quick-in-the-head position fixing are skills that are more difficult to teach, learn, and retain than those required for dead reckoning.

c. **Combination of Techniques.** Actually, the most successful navigation is obtained by combining the techniques described above. Constant orientation of the map and continuous observation of the terrain in conjunction with compass-read azimuths, and distance traveled on the ground compared with map distance, used together make reaching a destination more certain. One should not depend entirely on compass navigation or map navigation; either or both could be lost or destroyed.

11-7. NIGHT NAVIGATION

Darkness presents its own characteristics for land navigation because of limited or no visibility. However, the techniques and principles are the same that are used for day navigation. The success in nighttime land navigation depends on rehearsals during the planning phase before the movement, such as detailed analysis of the map to determine the type of terrain in which the navigation is going to take place and the predetermination of azimuths and distances. Night vision devices (see Appendix H) can greatly enhance night navigation.

a. The basic technique used for nighttime land navigation is dead reckoning with several compasses recommended. The point man will be in front of the navigator but just a few steps away for easy control of the azimuth. Smaller steps are taken during night navigation, so remember, the pace count will be different. It is recommended that a pace count obtained by using a predetermined 100-meter pace course be used at night.

b. Navigation using the stars is recommended in some areas; however, a thorough knowledge of constellations and location of stars is needed (see paragraph 9-5c). The four cardinal directions can also be obtained at night by using the same technique described for the shadow-tip method. Just use the moon instead of the sun. In this case, the moon has to be bright enough to cast a shadow.

CHAPTER 12

MOUNTED LAND NAVIGATION

A vehicle commander should be able to navigate from one point on the ground to another with or without a compass. If separated from his unit and given an azimuth and distance from their position to his, he should be able to reach the unit and continue the mission. To move effectively while mounted, he must know the principles of mounted navigation.

12-1. PRINCIPLES

The principles of land navigation while mounted are basically the same as while dismounted. The major difference is the speed of travel. Walking between two points may take one hour, but riding the same distance may only take 15 minutes. To be effective at mounted land navigation, the travel speed must be considered.

12-2. NAVIGATOR'S DUTIES

The duties of a navigator are so important and exacting that he should not be given any other duties. The leader should never try to be the navigator, since his normal responsibilities are heavy, and one or the other job would suffer.

a. **Assembling Equipment.** The navigator must gather all the equipment that will help him perform his job (maps, pencils, and so forth). He must do this before the mission starts.

b. **Servicing Equipment.** It is the navigator's duty to make sure that all the equipment he may use or require is working.

c. **Recording Data for Precise Locations.** During movement, the navigator must make sure that the correct direction and distance are recorded and followed. Grid coordinates of locations must be recorded and plotted.

d. **Supplying Data to Subordinate Leaders.** During movement, any change in direction or distance must be given to the subordinate leaders in sufficient time to allow them to react.

e. **Maintaining Liaison with the Commander.** The commander will normally select the route he desires to use. The navigator is responsible for following that route; however, there may be times when the route must be changed during a tactical operation. For this reason, the navigator must maintain constant communication with the commander. The navigator must inform the commander when checkpoints are reached, when a change in direction of movement is required, and how much distance is traveled.

12-3. MOVEMENT

When preparing to move, the effects of terrain on navigating mounted vehicles must be determined. You will cover great distances very quickly, and you must develop the ability to estimate the distance you have traveled. Remember that 0.1 mile is roughly 160 meters, and 1 mile is about 1,600 meters or 1.6 kilometers. Having a mobility advantage helps while navigating. Mobility makes it much easier if you get disoriented to move to a point where you can reorient yourself.

NOTE: To convert kilometers per hour to miles per hour, multiply by .62 (9 kph x .62 =5.58 mph). To convert miles per hour to kilometers per hour, divide miles per hour by .62 (10 mph . 0.62 = 16.12 kph)

a. **Consider Vehicle Capabilities.** When determining a route to be used when mounted, consider the capabilities of the vehicles to be used. Most military vehicles are limited in the degree of slope they can climb and the type of terrain they can negotiate. Swamps, thickly wooded areas, or deep streams may present no problems to dismounted soldiers, but the same terrain may completely stop mounted soldiers. The navigator must consider this when selecting a route

(1) Most vehicles will knock down a tree. The bigger the vehicle, the bigger the tree it can knock down. Vehicles cannot knock down several trees at once. It is best to find paths between trees that are wide enough for your vehicle. Military vehicles are designed to climb 60 percent slopes on a dry, firm surface (Figure 12-1, Page 12-2).

(2) You can easily determine approximate slope; just look at the route you have selected. If there is a contour line in any 100 meters of map distance on that route, it is a 10 percent slope. If there are two contour lines, it is 20 percent, and so forth. If there are four contour lines in any 100 meters, look for another route.

(3) Side slope is even more important than the slope you can climb. Normally, a 30 percent slope is the maximum in good weather. If you traverse a side slope, do it slowly and without turns. Rocks, stumps, or sharp

Figure 12-1. Tracked vehicle capabilities.

turns can cause you to throw the downhill track under the vehicle, which would mean a big recovery task.

(4) For tactical reasons, you will often want to move in draws or valleys because they give you cover. However, side slopes will force you to move slowly.

NOTE: The above figures are true for a 10-meter or a 20-foot contour interval. If the map has a different contour interval, just adjust the arithmetic. For instance, with one contour line in 100 meters, a 20-meter interval would give a 20 percent slope.

b. Know the Effects of Weather on Vehicle Movement. Weather can halt mounted movement. Snow and ice are obvious dangers, but more significant is the effect of rain and snow on soil load-bearing ability. Cross-country vehicles may be restricted to road movement in heavy rain. If it has rained recently, adjust your route to avoid flooded or muddy areas. A mired vehicle only hinders combat capability.

c. Prepare Before Movement. Locate the start point and finish point on the map. Determine the map's grid azimuth from start point to finish point and convert it to a magnetic azimuth. Determine the distance between the start point and finish point or any intermediate points on the map and make a thorough map reconnaissance of that area.

12-4. TERRAIN ASSOCIATION NAVIGATION

This is currently the most widely used method of navigation. The navigator plans his route so that he moves from terrain feature to terrain feature. An automobile driver in a city uses this technique as he moves along a street or series of streets, guiding on intersections or features such as stores and parks. Like the driver, the navigator selects routes or "streets"

between key points or "intersections." These routes must be capable of sustaining the travel of the vehicle or vehicles, should be relatively direct, and should be easy to follow. In a typical move, the navigator determines his location, determines the location of his objective, notes the position of both on his map, and then selects a route between the two. After examining the terrain, he adjusts the route by the following actions:

a. **Consider Tactical Aspects.** Avoid skylining, se-lect key terrain for overwatch positions, and select con-cealed routes.

b. **Consider Ease of Movement.** Use the easiest possible route and bypass difficult terrain. Remember that a difficult route will be harder to follow, be noisier, cause more wear and tear (and possible recovery problems), and take more time. Tactical surprise is achieved by doing the unexpected. Try to select an axis or corridor instead of a specific route Make sure you have enough maneuver room for the vehicles (Figure 12-2).

c. **Use Terrain Features as Checkpoints.** These must be easily recognizable in the light and weather conditions and at the speed at which you will move. You should be able to find a terrain feature from your location that can be recognized from almost anywhere and used as a guide. An example is checkpoint 2, the church, and checkpoint 3, the orchard, in Figure 12-2.

(1) The best checkpoints are linear features that cross your route. Use streams, rivers, hard-top roads, ridges, valleys, and railroads.

(2) The next best checkpoints are elevation changes, such as hills, depressions, spurs, and draws. Look for two contour lines of change. You will not be able to spot less than two lines of change while mounted.

(3) In wooded terrain, try to locate checkpoints at no more than 1,000-meter intervals. In open terrain, you may go to about 5,000 meters.

d. **Follow Terrain Features.** Movement and navigation along a

Figure 12-2. Primary route.

valley floor or near (not necessarily on) the crest of a ridgeline is easiest.

e. **Determine Directions.** Break the route down into smaller segments and determine the rough directions that will be followed. You do not need to use the compass; just use the main points of direction (north, northeast, east, and so forth). Before moving, note the location of the sun and locate north. Locate changes of direction, if any, at the checkpoints picked.

f. **Determine Distance.** Get the total distance to be traveled and the approximate distance between checkpoints. Plan to use the vehicle odometer to keep track of distance traveled. Use the pace count method and keep a record of the distance traveled. When using a pace count, convert from map distance to ground distance by adding the conversion factors of 20 percent for cross-country movement.

g. **Make Notes.** Mental notes are usually adequate. Try to imagine what the route will be like and remember it.

h. **Plan to Avoid Errors.** Restudy the route selected. Try to determine where errors are most apt to occur and how to avoid any trouble.

i. **Use a Logbook.** When the routes have been selected and the navigator has divided the distance to be traveled into legs, prepare a logbook. The logbook is an informal record of the distance and azimuth of each leg, with notes to aid the navigator in following the correct route. The notes list easily identifiable terrain feature at or near the point where the direction of movement changes (Figure 12-3).

12-5. DEAD RECKONING NAVIGATION

Dead reckoning is moving a set distance along a set line. Generally, it involves moving so many meters along a set line, usually an azimuth in degrees. There is no accurate method) of determining a direction in a moving vehicle. A magnetic vehicle-heading reference unit may be available in a few years; for now, use a compass.

a. **With Steering Marks.** This procedure is the same for vehicle travel as on foot.

(1) The navigator dismounts from the vehicle and moves away from the vehicle (at least 18 meters).

(2) He sets the azimuth on the compass and picks a steering mark (rock, tree, hilltop) in the direction on that azimuth (Figure 12-4).

(3) He remounts and has the driver identify the steering mark and proceeds to it in as straight a line as possible.

(4) On arrival at the steering mark or on any changes in direction, he repeats the first three steps above for the next leg of travel.

ODOMETER READING AT START	ODOMETER READING AT FINISH	DISTANCE IN MILES	AZIMUTH	DEVIATION CORRECTION	NOTES

Figure 12-3. Sample of a logbook format

Figure 12-4. Determining an azimuth, dismounted.

b. **Without Steering Marks.** This procedure is used only on flat, featureless terrain.

(1) The navigator dismounts from the vehicle, which is oriented in the direction of travel, and moves at least 18 meters to the front of the vehicle.

(2) He faces the vehicle and reads the azimuth to the vehicle. By adding or subtracting 180°, he determines the forward azimuth (direction of travel).

(3) On order from the navigator, the driver drives on a straight line to the navigator.

(4) The navigator remounts the vehicle, holds the compass as it will be held while the vehicle is moving, and reads the azimuth in the direction of travel.

(5) The compass will swing off the azimuth determined and pick up a constant deviation. For instance, say the azimuth was 75° while you were away from the vehicle. When you remounted and your driver drove straight forward, your compass showed 67°. You have a deviation of -8°. All you need to do is maintain that 67° compass heading to travel on a 75° magnetic heading.

(6) At night, the same technique can be used. From the map, determine the azimuth you are to travel. Convert the grid azimuth to a magnetic azimuth. Line the vehicle up on that azimuth, then move well in front of it. Be sure it is aligned correctly. Then mount, have the driver move slowly forward, and note the deviation. If the vehicle has a turret, the above procedure will work unless you traverse the turret; this will change the deviation.

(7) The distance factor in dead reckoning is easy. Just determine the map distance to travel and add 20 percent to convert to ground distance. Use your vehicle odometer to be sure you travel the proper distance.

12-6. STABILIZED TURRET ALIGNMENT NAVIGATION

Another method, if you have a vehicle with a stabilized turret, is to align the turret on the azimuth you wish to travel, then switch the turret stabilization system on. The gun tube will remain pointed at your destination no matter which way you turn the vehicle. This technique has been proven; it works. It is not harmful to the stabi-lization system. It is subject to stabilization drift, so use it for no more than 5,000 meters before resetting.

NOTE: If you have to take the turret off-line to engage a target, you will have to start all over, re-do the entire process.

12-7. COMBINATION NAVIGATION

Some mounted situations may call for you to combine and use both methods. Just remember the characteristics of each.

a. Terrain association is fast, is error-tolerant, and is best under most circumstances. It can be used day or night if you are proficient in it.

b. Dead reckoning is very accurate if you do everything correctly. You must be very precise. It is also slow, but it will work in very flat terrain.

c. You frequently will combine both. You may use dead reckoning to travel across a large, flat area to a ridge, then use terrain association for the rest of the move.

d. You must be able to use both methods. You should remember that your probable errors, in order of frequency, will be—

- Failure to determine distance(s) to be traveled.
- Failure to travel the proper distance.
- Failure to properly plot or locate the objective.
- Failure to select easily recognized check points or landmarks.
- Failure to consider the ease of movement factor.

NAVIGATION IN DIFFERENT TYPES OF TERRAIN

The information, concepts, and skills already presented will help you to navigate anywhere in the world; however, there are some special considerations and helpful hints that may assist you in various special environments. The following information not doctrine.

13-1. DESERT TERRAIN

About 5 percent of the earth's land surface is covered by deserts (Figure 13-1). Deserts are large arid areas with little or no rainfall during the year. There are three types of deserts—mountain, rocky plateau, and sandy or dune deserts. All types of forces can be deployed in the desert. Armor and mechanized infantry forces are especially suitable to desert combat except in rough mountainous terrain where light infantry may be required. Airborne, air assault, and motorized forces can also be advantageously employed to exploit the vast distances characteristic of desert warfare.

a. **Desert Regions.** In desert reagons, terrain varies from nearly flat to lava beds and salt marshes. Mountains deserts contain scattered ranges or areas of barren hills or mountains. The following are some of the world's major desert regions and their locations.

Region	Location
Sahara	North Africa
Kalahari	Southwest Africa
Arabian	Southwest Asia
Gobi	Mongolia and Northern China
Rub'al Khali	South Arabia
Great Basin, Colorado, Chihuahua, Yuma Sonoran, and Mohave	Northern Mexico and Western United States
Takla Makan	Northern China
Kyzyl Kum	Southwest USSR
Kara Kum	Southwest USSR
Syrian	Saudi Arabia, Jordan, and Iraq
Great Victoria	Western and South Australia
Great Sandy	Northwestern Australia
Patagonia	Southern Argentina and Chile
Atacama	Northern Chile

Figure 13-1 Deserts.

(1) Finding your way in a desert presents some degree of difficulty for a person who has never been exposed to this environment. Desert navigators have learned their way through generations of experience.

(2) Normally, desert people are nomadic, constantly moving in caravans. Navigating becomes second nature to them. Temperature in the tropical deserts reaches an average of 110° to 115° during the day, so most navigation takes place at night using the stars. Most deserts have some prevailing winds during the seasons. Such winds will arrange the sand dunes in a specific pattern that gives the navigator the opportunity to determine the four cardinal directions. He may also use the sun's shadow-tip method.

(3) A sense of direction also can be obtained by watching desert animals on their way to and from water holes (oases). Water, navigation, and survival are closely related in desert areas. Most deserts have pigeons or doves, and their drinking habits are important to the navigator. As a rule, these birds never drink in the morning or during the day, making their evening flights the most important. When returning from the oases, their bodies are heavier from drinking and their flight is accompanied by a louder flapping of their wings.

(4) Visibility is also an important factor in the desert, especially in judging distance. The absence of trees or other features prevents comparison between the horizon and the skyline.

b. **Interpretation and Analysis.** Many desert maps are inaccurate, which makes up-to-date air, aerial photo, and ground reconnaissance necessary In desert mountain areas contour intervals are generally large, so many of the intermediate relief features are not shown.

(1) The desert normally permits observation and fire to maximum ranges. The terrain is generally wide open and the exceptionally clear atmosphere offers excellent long-range visibility. Combine this with a powerful sun and low cloud density and you have nearly unlimited light and visual clarity, which often contribute to gross underestimations of ranges. Errors of up to 200 or 300 percent are not uncommon. However, visibility conditions may be severely affected by sandstorms and mirages (heat shimmer caused by air rising from the extremely hot daytime desert surface), especially if the observer is looking into the sun through magnifying optical instruments.

(2) Cover can be provided only by terrain feature masking because of the lack of heavy vegetation and man-made objects. It only takes a few meters of relief to provide cover. Concealment in the desert is related to the following six factors:

1. Shape. In order not to be observed by the enemy, attempt to alter the standard shapes of vehicles so they and their shadows are not instantly recognized.

2. Shine. Shine or glitter is often the first thing that attracts the observer's eye to movement many kilometers away. It must be eliminated.

3. Color and texture. All equipment should either be pattern painted or mudded to blend in with the terrain.

4. Light and noise. Light and noise discipline are essential because sound and light travel great distances in the desert.

5. Heat. Modern heat image technology makes shielding heat sources an important consideration when trying to hide from the enemy. This is especially important during night stops.

6. Movement. Movement itself creates a great deal of noise and dust, but a rapid execution using all the advantages the topography offers can help conceal it.

c. **Navigation.** When operating in the broad basins between mountain ranges or on rocky plateau deserts, there are frequently many terrain features to guide your movement by. But, observing these known features over great distances may provide a false sense of security in determining your precise location unless you frequently confirm your location by resection or referencing close-in terrain features. It is not uncommon to develop errors of several kilometers when casually estimating a position in this manner. Obviously, this can create many problems when attempting to locate a small checkpoint or objective, calling for CAS, reporting operational or intelligence information, or meeting CSS requirements.

(1) When operating in an area with few visual cues, such as in a sandy or dune desert, or when visibility is restricted by a sandstorm or darkness, you must proceed by dead reckoning. The four steps and two techniques for navigation presented earlier remain valid in the desert. However, your understanding of the special conditions found there will be extremely helpful as you apply them.

(2) Tactical mobility and speed are key to successful desert operations. Obstacles and areas such as lava beds or salt marshes, which preclude surface movements, do exist. But most deserts permit two-dimensional movement by ground forces similar to that of a naval task force at sea. Speed of execution is essential. Everyone moves farther and faster on the desert. Special navigation aids sometimes used in the desert include:

(a) Sun compass. It can be used on moving vehicles and sextants. It requires accurate timekeeping. However, the deviation on a magnetic compass that is caused by the metal and electronics in the vehicle is usually less than + 10°.

(b) Gyro compass. The gun azimuth stabilizer is in fact a gyro compass. If used on fairly flat ground, it is useful for maintaining direction over limited distances.

(c) Fires. Planned tracer fire or mortar and artillery concentrations (preferably smoke during the day and illumination at night) provide useful checks on estimated locations.

(d) Prepositioned lights. This method consists of placing two or more searchlights far apart, behind the line of contact, beyond enemy artillery range, and concealed from enemy ground observation. Units in the area can determine their own locations through resection, using the vertical beams of the lights. These lights must be moved on a time schedule known to all friendly units.

(3) One final note on desert navigation is that the sand, hard-baked ground, rocky surfaces, thorny vegetation, and heat generally found in the desert impose far greater demands for maintenance than you would plan for in temperate regions. It may also take longer to perform that maintenance.

13-2. MOUNTAIN TERRAIN

Mountains are generally understood to be larger than hills. Rarely do mountains occur individually; in most cases, they are found in elongated ranges or circular groups. When they are linked together, they constitute a mountain system (Figure 13-2). Light forces (infantry, airborne, and air assault forces) can operate effectively in mountainous regions because they are not terrain limited. Heavy forces must operate in passes and valleys that are negotiable by vehicle.

a. **Major Systems.** Some of the major systems include the following:

System	Location
The Andes	Central and South America
The Rockies	North America (USA-Canada)
The Appalachians	North America (USA-Canada)
The Alps	Central Europe
The Himalayas	Asia
The Caucasus	Western Asia and Europe (Russia)

b. **Minor Systems.** Some other systems are in Antarctica, Hawaii, Japan, New Zealand, and Oceania. Mountain systems are characterized by high, inaccessible peaks and steep slopes. Depending on the altitude, they may be snow covered. Prominent ridges and large valleys are also found. Navigating in this type of terrain is not difficult providing you make a careful examination of the map and the terrain.

c. **Climate.** Because of the elevations, it is always colder (3° to 5° per 300-meter gain in altitude) and

Figure 13-2. Mountain systems.

wetter than you might expect. Wind speeds can increase the effects of the cold even more. Sudden severe storms and fog are encountered regularly. Below the tree line, vegetation is heavy because of the extra rainfall and the fact that the land is rarely cleared for farming.

d. **Interpretation and Analysis.** The heights of mountainous terrain permit excellent long-range observation. However, rapidly fluctuating weather with frequent periods of high winds, rain, snow or fog may limit visibility. Also, the rugged nature of the terrain frequently produces significant dead space at mid-ranges.

(1) Reduced mobility, compartmented terrain, and the effects of rapidly changing weather increase the importance of air, ground, aerial photo, and map reconnaissance. Since mountain maps often use large contour intervals, microrelief interpretation and detailed terrain analysis require special emphasis.

(2) At first glance, some mountainous terrain may not appear to offer adequate cover and concealment; however, you can improve the situation. When moving, use rock outcroppings, boulders, and heavy vegetation for cover and concealment; use terrain features to mask maneuvers. Use harsh weather, which often obscures observation, to enhance concealment.

(3) Since there are only a few routing options, all-round security must be of primary concern. Natural obstacles are everywhere, and the enemy can easily construct more.

e. **Navigation.** Existing roads and trails offer the best routes for movement. Off-road movement may enhance security provided there is detailed reconnaissance, photo intelligence, or information from local inhabitants to ensure the route is negotiable. Again, the four steps and two techniques for navigation presented earlier remain valid in the mountains. Nevertheless, understanding the special conditions and the terrain will help you navigate. Other techniques that are sometimes helpful in mountains are:

(1) *Aspect of slope.* To determine the aspect of slope, take a compass reading along an imaginary line that runs straight down the slope. It should cut through each of the contour lines at about a 90° angle. By checking the map and knowing the direction of slope where you are located, you will be able to keep track of your location, and it will help guide your cross-country movement even when visibility is poor.

(2) *Use of an altimeter.* Employment of an altimeter with calibrations on the scale down to 10 or 20 meters is helpful to land navigators moving in areas where radical changes in elevation exist. An altimeter is a type of barometer that gauges air pressure except it measures on an adjustable scale marked in feet or meters

of elevation rather than in inches or centimeters of mercury. Careful use of the altimeter helps to pinpoint your position on a map through a unique type of resection. Instead of finding your position by using two different directional values, you use one directional value and one elevation value.

13-3. JUNGLE TERRAIN

These large geographic regions are found within the tropics near the equator (Central America, along the Amazon River, South-Eastern Asia and adjacent is lands, and vast areas in the middle of Africa and India) (Figure 13-3). Jungles are characterized as rainy, humid areas with heavy layers of tangled, impenetrable vegetation. Jungles contain many species of wildlife (tigers, monkeys, parrots, snakes, alligators, and so forth). The jungle is also a paradise for insects, which are the worst enemy of the navigator because some insects carry diseases (malaria, yellow fever, cholera, and so forth). While navigating in these areas, very little terrain association can be accomplished because of the heavy foliage. Dead reckoning is one of the methods used in these areas. A lost navigator in the jungle can eventually find his way back to civilization by following any body of water with a downstream flow. However, not every civilization found is of a friendly nature.

a. **Operations.** Operations in jungles tend to be isolated actions by small forces because of the difficulties encountered in moving and in maintaining contact between units. Divisions can move cross-country slowly; hut, aggressive reconnaissance, meticulous intelligence collection, and detailed coordination are required to concentrate forces in this way. More commonly, large forces operate along roads or natural avenues of movement, as was the case in the mountains. Patrolling and other surveillance operations are especially important to ensure security of larger forces in the close terrain of jungles.

(1) Short fields of observation and fire, and thick vegetation make maintaining contact with the enemy difficult. The same factors reduce the effectiveness of indirect fire and make jungle combat primarily a fight between infantry forces. Support by air and mechanized forces can be decisive at times, but it will not always be available or effective.

(2) Jungles are characterized by high temperatures, heavy rains, high humidity, and an abundance of vegetation. The climate varies with location. Close to the equator, all seasons are nearly alike with heavy rains all year. Farther from the equator (India and Southeast Asia), there are distinct wet (monsoon) and dry seasons. Both zones have high temperatures (averaging 75 to 95+ degrees Fahrenheit), heavy rainfall (as much as 400+ inches annually, and high humidity (90 percent) all year.

Figure 13-3. Jungles and savannas.

(3) In temperate climates, it is the areas of vegetation that are most likely to be altered and incorrectly portrayed on a map. In jungle areas, the vegetation grows so rapidly that it is more likely to be cleared and make these areas be shown incorrectly.

b. Interpretation and Analysis. The jungle environment includes dense forests, grasslands, swamps, and cultivated areas. Forests are classified as primary and secondary based upon the terrain and vegetation. Primary forests include tropical rain forests and deciduous forests. Secondary forests are found at the edges of both rain forests and deciduous forests and in areas where jungles have been cleared and abandoned. These places are especially overgrown with weeds, grasses, thorns, ferns, canes, and shrubs. Movement is especially slow and difficult. The extremely thick vegetation reaches a height of 2 meters and severely limits observation to only a few meters.

(1) Tropical rain forests consist mostly of large trees whose branches spread and lock together to form canopies. These canopies, which can exist at two and three different levels, may form as low as 10 meters from the ground. They prevent direct sunlight from reaching the ground, causing a lack of undergrowth on the jungle floor. Extensive above-ground root systems and hanging

vines are common and make vehicular travel difficult; foot movement is easier. Ground observation is limited to about 50 meters and air observation is nearly impossible.

(2) Deciduous forests are in semitropical zones that have both wet and dry seasons. In the wet season, trees are fully leaved; in the dry season, much of the folliage dies. Trees are usually less dense than in rain forests, which allows more sunlight to filter to the ground. This produces thick undergrowth. During the wet season, air and ground observations limited and movement is difficult. During the dry season, both improve.

(3) Swamps are common to all low, jungle areas where there is poor drainage. When navigating in a swampy area, a careful analysis of map and ground should be taken before any movement. The soldiers should travel in small numbers with only the equipment required for their mission, keeping in mind that they are going to be immersed in water part of the time. The usual technique used in swamp navigation is dead reckoning. There are two basic types of swamps— mangrove and palm. Mangrove swamps are found in coastal areas wherever tides influence water flow. Mangrove is a shrub-like tree that grows 1 to 5 meters high. These trees

have a tangled root system, both above and below the waterline, which restricts movement either by foot or small boat. Observation on the ground and from the air is poor but concealment is excellent.

(4) Grassy plains or savannas are generally located away from the equator but within the tropics. These vast land areas are characterized by flatlands with a different type of vegetation than jungles. They consist mainly of grasses (ranging from 1 to more than 12 feet in height), shrubs, and isolated trees. The most difficult areas to navigate are the ones surrounded by tall grass (elephant grass); however, vehicles can negotiate here better than in some areas. There are few or no natural features to navigate by, making dead reckoning or navigation by stars the only technique for movement (Figure 13-3, page 13-5). Depending on the height of the grass, ground observation may vary from poor to good. Concealment from air observation is poor for both soldiers and vehicles.

(5) Bamboo stands are common throughout the tropics. They should be bypassed whenever possible. They are formidable obstacles for vehicles, and soldier movement through them is slow, exhausting, and noisy.

(6) Cultivated areas exist in jungles also. They range from large, well-planned, well-managed farms and plantations to small tracts, cultivated by farmers. The three general types of cultivated areas are rice paddies, plantations, and small farms.

c. **Navigation.** Areas such as jungles are generally not accurately mapped because heavy vegetation makes aerial surveys difficult. The ability to observe terrain features, near or far, is extremely limited. The navigator must rely heavily upon his compass and the dead reckoning technique when moving in the jungle. Navigation is further complicated by the inability to make straight-line movements. Terrain analysis, constant use of the compass, and an accurate pace count are essential to navigation in this environment.

(1) Rates of movement and pace counts are particularly important to jungle navigators. The most common error is to overestimate the distance traveled. The distances below can be used as a rough guide for the maximum distances that might be traveled in various types of terrain during one hour in daylight.

Type of terrain	Maximum Distance (in Meters)
Tropical rain forest...................	up to 1,000
Deciduous forest........................	500
Secondary jungle.....................	100 to 500
Tall grass............................	500
Swamps............................	100 to 300
Rice paddies (wet)......................	800
Rice paddies (dry)......................	2,000
Plantations...........................	2,000
Trails............................	up to 3,000

(2) Special navigation strategies that are helpful in jungles include:

(a) Personal pace table. You should either make a mental or written personal pace table that includes your average pace count per 100 meters for each of the types of terrain through which you are likely to navigate.

(b) Resection using indirect fire. Call for mortar or artillery fire (airbursts of white phosphorous or illumination) on two widely separated grids that are not on terrain features like the one you are occupying and are a safe distance from your estimated location. Directions to the airbursts sometimes must be determined by sound.

(c) Modified area/point navigation. Even when making primary use of the compass for dead reckoning, you are frequently able to area navigate to an expanded objective, which is easily identified by terrain association. Then, simply develop a short, point-navigation leg to your final destination.

13-4. ARCTIC TERRAIN

Arctic terrain includes those areas that experience extended periods of below freezing temperatures. In these areas, the ground is generally covered with ice or snow during the winter season. Although frozen ground and ice can improve trafficability, a deep accumulation of snow can reduce it. Vehicles and personnel require special equipment and care under these adverse conditions.

a. **Operations.** Both the terrain and the type and size of unit operations will vary greatly in arctic areas. In open terrain, armored and mechanized forces will be effective although they will have to plan and train for the special conditions. In broken terrain, forests, and mountains, light forces will predominate as usual. However, foot movement can take up to five times as long as it might under warmer conditions.

b. **Interpretation and Analysis.** Both the terrain and cultural features you may confront in winter may vary to any extreme, as can the weather. The common factor is an extended period of below-freezing temperatures. The terrain may be plains, plateaus, hills, or mountains. The climate will be cold, but the weather will vary greatly from place to place. Most arctic terrain experiences snow, but some claim impressive accumulations each season, such as the lake-effected snow belts off Lake Ontario near Fort Drum, New York. Other areas have many cold days with sunshine and clear nights, and little snow accumulation.

(1) In areas with distinct local relief and scattered trees or forests, the absence of foliage makes movement by terrain association easier; observation and fields of fire are greatly enhanced except during snowstorms. But in relatively flat, open areas covered with snow (especially in bright sunlight), the resulting lack of contrast

may interfere with your being able to read the land. With foliage gone, concealment (both from the ground and from the air) is greatly reduced. As in desert areas, you must make better use of the terrain to conceal your movements.

(2) Frozen streams and swamps may no longer be obstacles, and thus identification of avenues of approach may be difficult in winter. However, the concept as to what is key terrain is not likely to be affected.

c. **Navigation.** Special skills may be required in arctic terrain, such as the proper use of winter clothing, skis, and snowshoes; but this does not affect your navigation strategies. There are no special techniques for navigating in arctic terrain. Just be aware of the advantages and disadvantages that may present themselves and make the most of your opportunities while applying the four steps and two techniques for land navigation.

(1) Remember, the highest caliber of leadership is required to ensure that all necessary tasks are performed, that security is maintained, and that soldiers and their equipment are protected from the physical effects of very low temperatures. There is a great temptation to do less than a thorough job at whatever the task may be when you are very cold.

(2) Night navigation may be particularly enhanced when operating in arctic terrain. Moonlight and starlight on a clear night reflect off the snow, thus enabling you to employ daytime terrain association techniques with little difficulty. Even cloudy winter nights are often brighter than clear moonlit summer nights when the ground is dark and covered with foliage. Movements with complete light discipline (no black-out drives) can often be executed. On the other hand, areas with severe winter climates experience lengthy periods of darkness each day, which may be accompanied by driving snow and limited visibility.

13-5. URBAN AREAS

The world continues to become more urbanized each year; therefore, it is unlikely that all fighting will be done in rural settings. Major urban areas represent the power and wealth of a particular country in the form of industrial bases, transportation complexes, economic institutions, and political and cultural centers. Therefore, it may be necessary to secure and neutralize them. When navigating in urban places, it is man-made features, such as roads, railroads, bridges, and buildings that become important, while terrain and vegetation become less useful.

a. **Interpretation and Analysis.** Military operations on urbanized terrain require detailed planning that provides for decentralized execution. As a result of the rapid growth and changes occurring in many urban areas, the military topographic map is likely to be outdated. Supplemental use of commercially produced city maps may be helpful, or an up-to-date sketch can be made.

(1) Urbanized terrain normally offers many AAs for mounted maneuver well forward of and leading to urban centers. In the proximity of these built-up areas, however, such approach routes generally become choked by urban sprawl and perhaps by the nature of adjacent natural terrain. Dismounted forces then make the most of available cover by moving through buildings and underground systems, along edges of streets, and over rooftops. Urban areas tend to separate and isolate units, requiring the small-unit leader to take the initiative and demonstrate his skill in order to prevail.

(2) The urban condition of an area creates many obstacles, and the destruction of many buildings and bridges as combat power is applied during a battle further limits your freedom of movement. Cover and concealment are plentiful, but observation and fields of fire are greatly restricted.

b. **Navigation.** Navigation in urban areas can be confusing, but there are often many cues that will present themselves as you proceed. They include streets and street signs; building styles and sizes; the urban geography of industrial, warehousing, residential housing, and market districts; man-made transportation features other than streets and roads (rail and trolley lines); and the terrain features and hydrographic features located within the built-up area. Strategies for staying on the route in an urban area include:

(1) *Process rouse descriptions.* Write down or memorize the route through an urban area as a step-by-step process. For example, "Go three blocks north, turn left (west) on a wide divided boulevard until you go over a river bridge. Turn right (north) along the west bank of the river, and. . ."

(2) *Conceptual understandings of the urban area.* While studying the map and operating in a built-up area, work hard to develop an understanding (mental map) of the entire area. This advantage will allow you to navigate over multiple routes to any location. It will also preclude your getting lost whenever you miss a turn or are forced off the planned route by obstacles or the tactical situation.

(3) *Resection.* Whenever you have a vantage point to two or more known features portrayed on the map, do not hesitate to use either estimated or plotted resection to pinpoint your position. These opportunities are often plentiful in an urban setting.

UNIT SUSTAINMENT

Land navigation is a skill that is highly perishable. The soldier must continually make use of the skills he has acquired to remain proficient in them. The institution is responsible for instruction in the basic techniques of land navigation. The institution tests these skills each time a soldier attends a leadership course. However, it is the unit's responsibility to develop a program to maintain proficiency in these skills between institution courses. The unit sustainment program provides training that will build on and reinforce the skills the soldier learned in the institution. It should use the building-block approach to training: basic map reading instruction or review, instruction on land navigation skills, dead reckoning training, dead reckoning practice, terrain association training, terrain association practice, land navigation testing, and building of leader skills. These reader skills should include following a route selected by the commander and planning and following a route selected by the leader. The unit trainer should be able to set up a sustainment program, a train-the-trainer program, and a land navigation course for his unit's use. It is recommended that units develop a program similar to the one outlined in this chapter. Complete lesson outlines and training plans are available by writing to Commander, 29th Infantry Regiment, ATTN: ATSHINB-A, Fort Benning, GA 319055595.

14-1. SET UP A SUSTAINMENT PROGRAM

The purpose of setting up a sustainment program in the unit is to provide soldiers with training that will reinforce and build on the training they have received in the institution. All soldiers should receive this training at least twice a year. It will also provide the unit with means of identifying the areas in which the soldiers need additional training.

a. **Training Guidance.** The unit commander must first determine the levels of proficiency and problems that his unit has in land navigation. This can be done through after-action reports from the unit's rotations to NTC/JRTC, ARTEP final reports, feedback from his subordinates, personal observation, and annual training. Once the unit commander decides where his training time should be concentrated, he can issue his training guidance to his subordinate leaders. He will also direct his staff to provide training sites, resources, and time for the units to train land navigation. It is recommended that land navigation be trained separately, not just included as a subtask in tactical training.

b. **Certification.** The unit commander must also provide his subordinate commanders with a means of certifying training. The unit staff must provide subject matter experts to ensure training meets the standards decided upon by the unit commander. Instructors should be certified to instruct, and courses should be certified prior to use by the unit.

c. **Program Development.** The sustainment program should be able to meet the requirements of all of the unit's soldiers. It should address all skills front basic map reading to leaders, planning and executing a route.

The program should cover the following:

- Diagnostic examination.
- Map reading instruction/review.
- Land navigation skills training.
- Dead reckoning training/practice.
- Terrain association training/practice.
- Land navigation written/field examination.
- Leaders' training and testing.

The sustainment program should be developed and th maintained in the unit's training files. The progra should be developed in training modules so that it can used as a whole program or used separately by individu modules. It should be designed so the commander c decide which training modules he will use, depending the proficiency of the unit. The unit commander ne only use those modules that fit his training plan.

14-2. SET UP A TRAIN-THE-TRAINER PROGRA

The purpose of a train-the-trainer program in the unit to develop trainers capable of providing soldiers with confidence and skills necessary to accomplish assigned land navigation tasks.

a. **Development of the Program.** The u commander should appoint a cadre of officers and NC to act as primary and alternate instructors for la navigation training. Use the training modules the u has developed and have these soldiers go through ea module of training until they can demonstrate experti Determine which instructors will conduct each module training and have them practice until they are fully

ared to give the training. These instructors act as
ning cadre for the entire unit. They train their peers
nstruct the subordinate units, and they certify each
t's training.

b. **Conduct of Training.** Conduct training at the
est level possible. Leaders must be included in all
ning to keep unit integrity intact.

3. SET UP AND NAVIGATION COURSE

e unit commander provides specific guidance on what
requires in the development of a land navigation
rse. It depends upon the unit's mission, training plan,
tasks to be trained. There are basic guidelines to use
en setting up a course.

a. **Determine the Standards.** The unit commander
ermines the standards for the course. Recommended
dards are as follows:

(1) Distance between points: no less than 300 me-
s; no more than 1,200 meters.

(2) Total distance of lanes: no less than 2,700
ters; no more than 11,000 meters.

(3) Total number of position stakes: no less than
en per lane; no more than nine per lane.

(4) Time allowed: no less than three hours; no more
n four hours.

b. **Decide on the Terrain.** The unit should use
terrain that is similar to terrain they will be using in
tactical exercises. Terrain should be different each time
training is conducted; the training area for a dismounted
course needs to be at least 25 square kilometers.
Mounted courses require twice as much terrain so that
vehicles are not too close to each other.

c. **Perform a Map and Ground Reconnaissance.**
Check the terrain to determine position stake locations,
look for hazards, and to develop training briefings.

(1) Plot the locations of your position stakes on a
1:50,000-scale map.

(2) Fabricate or order position stakes.

(3) Request support from the local engineer or field
artillery unit to survey the position stakes in.

(4) Survey the position stakes in and emplace them.

(5) Certify the course by having your SMEs negotiate each lane of the course.

(6) Prepare course requirement sheets and print
them.

(7) Complete a risk assessment of the training area.

(8) Begin teaching.

This sequence can be used to develop any type of land
navigation course. The difference in courses will depend
on the commander's guidance.

FIELD SKETCHING

*A sketch is a free-hand drawing of a map or picture of an area or route of travel. It shows enough
detail and has enough accuracy to satisfy special tactical or administrative requirements.*

A-1. PURPOSE

Sketches are useful when maps are not available or the
existing maps are not adequate, or to illustrate a reconnaissance or patrol report. Sketches may vary from
hasty to complete and detailed, depending upon their
purpose and the degree of accuracy required. For example,
a sketch of a large minefield will require more accuracy
than a hasty sketch of a small unit's defensive position.

A-2. MILITARY SKETCHES

The scale of a sketch is determined by the object in view
and the amount of detail required to be shown. The sketch
of a defensive position for a platoon or company will
normally call for a sketch of larger scale than a sketch for
the same purpose for a division. A field sketch (Figure A-
1) must show the north arrow, scale, legend, and the
following features:

- Power lines.
- Rivers.
- Main roads.
- Towns and villages.
- Forests.
- Rail lines.
- Major terrain features.

Military sketches include road and area sketches.

a. Road sketches show the natural and military f
tures on and in the immediate vicinity of the road.
general, the width of terrain sketches will not exceed 3
meters on each side of the road. Road sketches may
used to illustrate a road when the existing map does
show sufficient detail.

b. Area sketches include those of positions, OPs,
particular places.

(1) *Position sketch.* A position sketch is one of
military position, campsite, or other area of ground,
effectively complete a position sketch, the sketcher m
have access to all parts of the area being sketched.

(2) *Observation post sketch.* An OP sketch shows
military features of ground along a friendly OP line as
toward the enemy position as possible.

(3) *Place sketch.* A place sketch is one of an a
made by a sketcher from a single point of observati
Such a sketch may cover ground in front of an OP line,
it may serve to extend a position or road sketch toward
enemy.

Figure A-1. Sketch map.

MAP FOLDING TECHNIQUES

One of the first considerations in the care of maps is its proper folding.

1. FOLDING METHODS

Figures B-1 and B-2 show ways of folding maps to make them small enough to be carried easily and still be available for use without having to unfold them entirely.

2. PROTECTION METHOD

After a map has been folded, it should be pasted in a folder for protection. Apply adhesive to the back of the segments corresponding to A, F, L, and Q (Figure B-2).

3. PRACTICE CUT

It is suggested that before attempting to cut and fold a map in the manner illustrated in Figure B-2, make a practice cut and fold with a piece of paper.

Figure B-2. Two methods of folding a map.

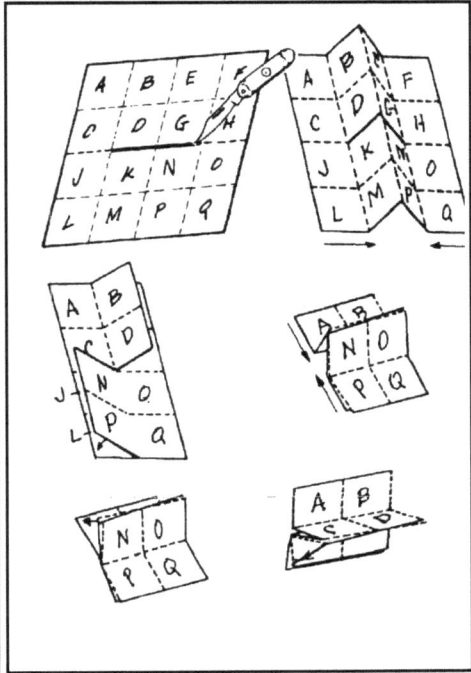

Figure B-1. How to slit and fold a map for special use.

UNITS OF MEASURE AND CONVERSION FACTORS

ENGLISH SYSTEM OF LINEAR MEASURE

12 inches	=	1 foot
36 inches	=	1 yard
3 feet	=	1 yard
1,760 yards	=	1 mile statute
2,026.8 yards	=	1 mile nautical
5,280 feet	=	1 mile statute
6,080.4 feet	=	1 mile nautical
63,360 inches	=	1 mile statute
72,963 inches	=	1 mile nautical

METRIC SYSTEM OF LINEAR MEASURE

1 millimeter	=	0.1 centimeter	=	0.0393 inch
10 millimeters	=	1.0 centimeter	=	0.3937 inch
10 centimeters	=	1.0 decimeter	=	3.937 inches
10 decimeters	=	1.0 meter	=	39.37 inches
10 meters	=	1.0 decameter	=	32.81 feet
10 decameters	=	1.0 hectometer	=	328.1 feet
10 hectometers	=	1.0 kilometer	=	0.62 mile
10 kilometers	=	1.0 myriameter	=	6.21 miles

EQUIVALENT UNITS OF ANGULAR MEASURE

1 mil	=	1/6400 circle	=	0.05625°	=	0.0625 grad
1 grad	=	1/400 circle	=	16.0 mils	=	0°54' = 0.9°
1 degree	=	1/360 circle	=	about 17.8mils	=	about 1.1 grad

CONVERSION FACTORS

One	Inches	Feet	Yards	Statute Miles	Nautical Miles	mm
Inch............................1	0.0833	0.0277				25.40
Foot...........................12	1	0.333				304.8
Yard...........................36	3	1	0.00056			914.4
Statute Mile............63,360	5,280	1,760	1	0.8684		
Nautical Mile...........72,963	6,080	2,026	1.1516	1		
Millimeter.....................0.0394	0.0033	0.0011				1
Centimeter....................0.3937	0.0328	0.0109				10
Decimeter.....................3.937	0.328	0.1093				1,00
Meter..........................39.37	3.2808	1.0936	0.0006	0.0005		1,000
Decameter....................393.7	32.81	10.94	0.0062	0.0054		10,000
Hactometer.................3,937	328.1	109.4	0.0621	0.0539		100,000
Kilometer................39,370	3,281	1,094	0.6214	0.5396		1,000,000
Myriameter............393,700	32,808	10,936	6.2137	5.3959		10,000,000

One	cm	dm	m	dkm	hm	km	mym
Inch................................2.540	0.2540	0.0254	0.0025	0.0003			
Foot 30.48	3.048	0.3048	0.0305	0.0030	0.0003		
Yard 91.44	9.144	0.9144	0.0914	0.0091	0.0009		
Statute Mile...............160,930	16,093	1,609	160.9	16.09	1.6093	0.1609	
Nautical Mile 185,325	18,532	1,853	185.3	18.53	1.8532	0.1853	
Millimeter 0.1	0.01	0.001	0.0001				
Centimeter...... 1	0.1	0.01	0.001	0.0001			
Decimeter 10	1	0.1	0.01	0.001	0.0001		
Meter 100	10	1	0.1	0.01	0.001	0.0001	
Decameter 1000	100	10	1	0.01	0.01	0.001	
Hactometer 10,000	1000	100	10	1	0.1	0.01	
Kilometer........ 100,000	10,000	1000	100	10	1	0.1	
Myriameter 1,000,000	100,000	10,000	1000	100	10	1	

Example I

Problem: Reduce 76 centimeters to (?) inches.

76 cm x 0.3937 = 29 inches

Answer: There are 29 inches in 76 centimeters.

Example II

Problem: How many feet are there in 2.74 meters?

$$\frac{274}{.3048} = 9 \text{ feet}$$

Answer: There are approximately 9 feet in 2.74 meters.

GROUND DISTANCE AT MAP SCALE

Scale	1 Inch Equals	1 Centimeter Equals
1:5,000	416.67 feet	164.0 feet
	127.00 meters	50.00 meters
1:10,000	833.33 feet	328.1 feet
	254.00 meters	100.00 meters
1:12,500	1,041.66 feet	410.1 feet
	317.00 feet	125.00 meters
1:20,000	1,666.7 feet	656.2 feet
	508.00 meters	200.00 meters
1:25,000	2,083.3 feet	820.2 feet
	635.00 meters	250.00 meters
1:50,000	4,166.7 feet	1,640.4 feet
	1,270.00 meters	500.00 meters
1:63,360	5,280.0 feet (1 mile)	2,078.7 feet
	1,609.30 meters	633.60 meters
1:100,000	8,333.3 feet	3,280.8 feet
	2,540.00 meters	1,000.00 meters
1:250,000	20,833.0 feet	8,202.0 feet
	6,350.00 meters	2,500.00 meters
1:500,000	41,667.0 feet	16,404.0 feet
	12,700.00 meters	5,000.00 meters

JOINT OPERATIONS GRAPHICS

Joint operations graphics (paragraph 2-6b[6]) are based on the format of the standard 1:250,000-scale military topographic maps. They contain additional information needed in present-day joint air-ground operations.

D-1. TYPES
Each JOG is prepared in two types; one is designed for air operations and the other for ground operations. Each version is identified in the lower margin as JOINT OPERATIONS GRAPHIC (AIR) or JOINT OPERATIONS GRAPHIC (GROUND).

D-2. BASIC CONTENTS
The basic topographic information is the same on both JOG versions.

a. Power transmission lines are symbolized as a series of purple pylons connected by a solid purple line.

b. Airports, landing facilities, and related air information are shown in purple. The purple symbols that may be unfamiliar to the user are shown in the legend in the margin.

c. The top of each obstruction to air navigation is identified by its elevation above sea level and its elevation above ground level.

d. Along the north and east edges of the graphic, detail is extended beyond the standard sheet lines to create an overlap with the graphics to the north and to the east.

e. Layer tinting (paragraph 10-1a) and relief shading (paragraph 10-1c) are added as an aid to interpreting the relief.

f. The incidence of the graphic in the world geographic reference system (paragraph 4-8b) is shown by a diagram in the margin.

D-3. JOINT OPERATIONS GRAPHIC (AIR)
The JOG (AIR) series, prepared for air use, contains detailed information on air facilities such as radio ranges, runway lengths, and landing surfaces. The highest terrain elevation in each 15-minute quadrangle is identified by the large open-faced figures shown in the legend. Elevations and contours on JOG (AIR) sheets are given in feet.

D-4. JOINT OPERATIONS GRAPHIC (GROUND)
The JOG (GROUND) series is prepared for use by ground units, and only stable or permanent air facilities are identified. Elevations and contours are located in the same positions as on the air version, but are given in meters.

APPENDIX E
EXPORTABLE TRAINING MATERIAL

This appendix provides information on the exportable training material available for unit training in basic land navigation skills. These materials are available from the Army Research Institute, Fort Benning Field Unit, Fort Benning, Georgia 31905.

E-1. PLANNING TO NAVIGATE

This training material describes the planning process in detail. Planning may be the most important aspect of land navigation. Planning to navigate includes background information for map interpretation, a practical example in planning to navigate, and a description of how to train these skills.

E-2. HOW TO TRAIN BASIC LAND NAVIGATION SKILLS

This training material offers guidance for unit training in basic land navigation skills. It includes training modules for distance and location skills. Each module is self-contained, so training can be given in each skill separately from other skills.

E-3. DISMOUNTED LAND NAVIGATION TECHNIQUE

This training material describes the process of navigating with an emphasis on movement skills and the techniques and strategies that should be used while navigating. Critical training covers how to put skills together for movement and how to decide which technique to use in a particular situation.

E-4. MAP INTERPRETATION TERRAIN ASSOCIATION COURSE

The MITAC program is designed to teach terrain association through a "building block" approach starting with simple elements first, then adding more complex information as the soldier progresses from one level to another. MITAC consists of three systematic courses of instruction: basic, intermediate, and advanced.

E-5. ROUTE PLANNING GUIDE

The Route Planning Guide provides the small-unit leader with a comprehensive reference document, which he can use in learning to plan dismounted administrative or tactical moves. It offers him a planning procedure that he can use in the field without any notes to plan successful moves over unfamiliar terrain.

E-6. LAND NAVIGATION SUSTAINMENT PROGRAM

The Land Navigation Sustainment Program is designed to develop trainers that are capable of providing soldiers with the confidence and skills necessary to accomplish all assigned land navigation tasks and therefore to develop soldiers capable of accomplishing these tasks.

ORIENTEERING

What is orienteering? Orienteering is a competitive form of land navigation It is for all ages and degrees of fitness and skill. It provides the suspense and excitement of a treasure hunt. The object of orienteering is to locate control points by using a map and compass to navigate through the woods. The courses may be as long as 10 kilometers.

F-1. HISTORY

Orienteering began in Scandinavia in the nineteenth century. It was primarily a military event and was part of military training. It was not until 1919 that the modern version of orienteering was born in Sweden as a competitive sport. Ernst Killander, its creator, can be rightfully called the father of orienteering. In the early thirties, the sport received a technical boost with the invention of a new compass, more precise and faster to use. The Kjellstrom brothers, Bjorn and Alvan, and their friend, Brunnar Tillander, were responsible for this new compass. They were among the best Swedish orienteers of the thirties, with several individual championships among them. Orienteering was brought into the United States in 1946 by Bjorn Kjellstrom.

F-2. DESCRIPTION

Each orienteer is given a 1:50,000 topographic map with the various control points circled. Each point has a flag marker and a distinctive punch that is used to mark the scorecard. Competitive orienteering involves running from checkpoint to checkpoint. It is more demanding than road running, not only because of the terrain, but because the orienteer must constantly concentrate, make decisions, and keep track of the distance covered. Orienteering challenges both the mind and the body; however, the competitor's ability to think under pressure and make wise decisions is more important than speed or endurance.

F-3. THE COURSE

The orienteering area should be on terrain that is heavily wooded, preferably uninhabited, and difficult enough to suit different levels of competition. The area must be accessible to competitors and its use must be coordinated with appropriate terrain and range control offices.

a. The ideal map for an orienteering course is a multi-colored, accurate, large-scale topographic map. A topographic map is a graphic representation of selected man-made and natural features of a part of the earth's surface plotted to a definite scale. The distinguishing characteristic of a topographic map is the portrayal of the shape and elevation of the terrain by contour lines.

b. For orienteering within the United States, large-scale topographic (topo) maps are available from the Defense Mapping Agency Hydrographic Topographic Center. The scale suitable for orienteering is 1:50,000 (DMA).

F-4. SETTING UP THE COURSE

The challenge for the course setter is to keep the course interesting, but never beyond the individual's or group's ability. General guidance is to select locations that are easily identifiable on the map and terrain, and accessible from several routes.

a. Those who set up the initial event should study a map for likely locations of control points and verification of the locations. Better yet, they should coordinate with an experienced competitor in selecting the course.

b. There are several forms of orienteering events. Some of the most common are route, line, cross-country, and score orienteering.

(1) *Route orienteering*. This form can be used during the training phase and in advanced orienteering. In this type of event, a master or advanced competitor leads the group as they walk a route. The beginners trace the actual route walked on the ground on their maps. They circle the location of the different control points found along the walked route. When they finish, the maps are analyzed and compared. During training, time is not a factor. Another variation is when a course is laid out on the ground with markers for the competitor to follow. There is no master map, as the course is traced for the competitor by flags or markers. The winner of the event is the competitor who has successfully traced the route and accurately plotted the most control points on his map.

(2) *Line orienteering*. At least five control points are used during this form of orienteering training. The competitor traces on his map a preselected route from a master map. The object is to walk the route shown on the map, circling the control points on the map as they are located on the ground (Figure F-1, Page F-2).

Figure F-1. Line Orienteering

(3)*Cross-country orienteering.* This is the most common type of orienteering competitions. It is sometimes called free or point orienteering and is considered to be the most competitive and intriguing of all events (Figure F-2). In this event, all competitors must visit the same controls in the same order. With the normal one-minute starting interval, it becomes a contest of route choice and physical skill. The winner is the contestant with the fastest time around the course.

(a) After selecting the control points for the course, deter-mine the start and finish location(s). The last control should be near the finish. In de-scribing each control's location, an eight-digit grid coordinate and a combination of two letters

Control Number	Coordinates	Control Code	Description of Clues
1	GL01589334	WE	Cemetery,northern part
2	GL02709323	WT	Creek junction-center
3	GL03509260	DA	Draw, center
4	GL02229155	WK	South end of runway
5	GL01709046	CM	East end of pond
6	GL00659125	RD	Hilltop 134-top
7	GL01509218	JD	Spur-150 southeast of hilltop

Figure F-2. A cross-country orienteering map.

identifying the point (control code) should be included in each descriptive clue list that is normally given to each competitor at least two minutes before his start time.

(b) There are usually 6 to 12 control markers on the course in varying degrees of difficulty and distances apart so that there are no easy, direct routes. Instead, each competitor is faced with many choices of direct but difficult routes, or of indirect but easier routes. Each control's location is circled, and the order in which each is to tee visited is clearly marked on the master map. The course may be a closed transverse with start and finish collocated, or the start and finish may be at different locations. The length of the course and difficulty of control placement will vary with the competitors' degree of expertise. Regardless of the class of event, all competitors must indicate on their event cards proof of visiting

the control markers. This is usually done by inked stamps, coded letters, or punches.

NOTE: The same orienteering range may serve in both cross-country and score events. However, a separate set of competitor maps, master maps, and event cards is necessary.

(4) *Score orienteering* In this event, the area chosen for the competition is blanketed with many control points (Figure F-3). The controls near the start/finish point (usually identical in this event) have a low point value, while those more distant or more difficult to locate have a high point value. (See Figure F-7 for a sample card.) This event requires the competitor to locate as many control markers as he can within the specified time (usually 90 minutes). Points are awarded for each control visited and deducted for exceeding the

Control Number	Coordinates	Control Code	Description of Clues
1	GL06938085	TK	Railroad bridge-north
2	GL08608218	RZ	Top of hill
3	GL06508535	CD	North of pond-swamp
4	GL09608265	PR	Center-creek junction
5	GL09858109	AC	Building-dead end
6	GL10968245	SM	Ruins
7	GL09208430	WN	Road Junction-north
8	GL06338275	YV	Building-west end of pond
9	GL07858050	NT	Top of hill
10	GL08158350	YM	East end of lake

Figure F-3. A score orienteering map

Figure F-4. Control markers.

specified time. The competitor with the highest point score is the winner.

(a) Conducting a score event at the start is basically the same as the cross-country event. The competitor is given a map and an event card. The event card lists all the controls with their different point values. When released to the master map, the competitor finds the circles and numbers indicating the location of all the controls listed on his event card. He copies all the red circles on his map. Then he chooses any route he wishes to take in amassing the highest possible point score in the time available. The course is designed to ensure that there

RECORDER'S SHEET							
NAME	No.	Team-Class	Start Time	Finish Time	Time Taken	Controls Missed	Position

Figure F-5. Recorder's sheet.

are more control points than can possibly be visited in the allotted time. Again, each control marker visited must be indicated on the event card.

(b) It is important for the competitor to take time initially to plot the most productive route. A good competitor may spend up to 6 minutes in the master map area while plotting the ideal route.

(c) There is no reward for returning early with time still available to find more points, so the good competitor must be able to coordinate time and distance with his ability in land navigation in running the course.

F-5. OFFICIALS

The same officials can be used at the start and finish. More officials or assistants can be used; the following material lists the minimum that can be used for a competition. They include the following:

a. **At The Start.**

(1) *Course organizer* Briefs orienteers in the assembly area, issues event cards and maps, and calls orienteers forward to start individually.

(2) *Recorder*—Records orienteer's name and start time on recorder's sheet, checks orienteer's name and start number on his event card, and issues any last-minute instructions.

(3) *Timer*—Controls the master clock and releases the orienteers across the start line at their start time (usually at one-minute intervals) to the master map area.

b. **At The Finish.**

(1) *Timer*—Records finish time of each orienteer on the orienteer's event card and passes card to recorder.

(2) *Recorder*—Records finish time of each orienteer on the orienteer's event card and passes card to recorder.

(3) *Course organizer*—Verifies correctness of names, finish times, and final score; posts orienteers' positions on results board; and accounts for all orienteers at the end of event.

F-6. START/FINISEI AREA

The layout of the start/finish areas for orienteering events is basically the same for all forms.

a. **Assembly Area.** This is where orienteers register and receive instructions, maps, event cards, and start numbers. They may also change into their orienteering clothes if facilities are available, study their maps, and fill out their event cards here. Sanitation facilities should be available in this area.

b. **Start.** At the start, the orienteer will report to the recorder and timer's table to be logged in by the recorder and released by the timer.

c. **Master Map Area.** There are three to five master maps 20 to 50 meters from the start. When the orienteer arrives at this area, he must mark his map with all the course's control points. Having done this, he must decide on the route he is to follow. The good orienteer will take the time to orient his map and carefully plot his route before rushing off. It is a good idea to locate the master map area out of sight of the start point to preclude orienteers tracking one another.

d. **Equipment.** The following is a list of equipment needed by the host of an orienteering event:

- Master maps, three to five, mounted.
- Competitor maps, one each.
- Event cards, one each.
- Recorder's sheets, two.
- Descriptive clue cards, one each.
- Time clocks, two.
- Rope, 100 to 150 feet, with pegs for finish tunnel.
- Card tables, one or two.
- Folding chairs, two or three.
- Results board.
- Control markers, one per point.
- Extra compasses.
- Whistle, for starting.
- First aid kit.
- Colored tape or ribbon for marking route to master map and from last control point to finish.

e. **Control Markers.** These are orange-and-white markers designating each control point (Figure F-4). Ideally, they should have three vertical square faces, forming a triangle with the top and bottom edges. Each face should be 12 inches on a side and divided diagonally into red and white halves or cylinders (of similar size) with a large, white, diagonal stripe dividing the red cylinder. For economy or expediency, 1-gallon milk cartons, 5-gallon ice cream tubs, 1-gallon plastic bleach bottles, or foot-square plagues, painted in the diagonal or divided red and white colors of orienteering, may be used.

(1) Each marker should have a marking or identification device for the orienteer to use to indicate his visit to the control. This marker may be the European-style punch pliers, a self-inking marker, different colored crayons at each point, different letter combinations, different number combinations, or different stamps or coupons. The marking device must be unique, simple, and readily transcribable to the orienteers' event cards.

(2) The control marker should normally be visible from at least 10 meters. It should not be hidden.

f. **Recorder's Sheets.** A suggested format for the recorder's sheet is depicted in Figure F-5.

g. **Event Card.** The event card can be made before the event and should be as small as possible, as it is carried by the competitor. It must contain the following items: name, start number, start time, finish time, total time, place, and enough blocks for marking the control points. As indicated earlier, it may also contain a listing of descriptive clues (Figure F-6).

CROSS-COUNTRY ORIENTEERING TEAM

NAME _____ COMPANY _____ COURSE _____ TEAM _____

NAME _____ COMPANY _____ START TIME _____ FINISH TIME _____

CHECKPOINTS   DESCRIPTION CLUES

1	2
3	4
5	6

NOTE: All control signes are located at eye level on trees.

Total Value of Points _____

Penalty Points _____

Final Score _____

1. All work is individual team effort.
2. You must not join with or coordinate with any other team.
3. You must personally visit each point you indicate on your scorecard.

Figure F-6. Cross-country orienteering event card.

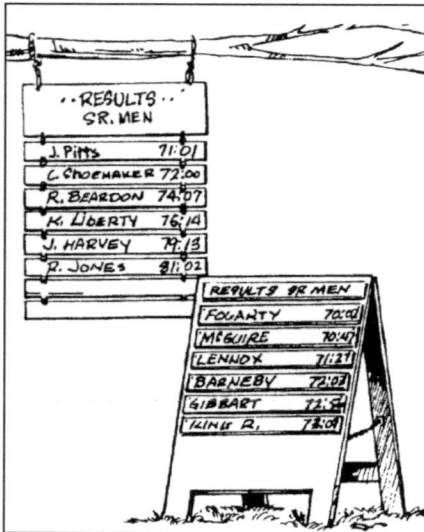

Figure F-7. Results board.

h. **Results Board.** This board displays the orienteer's position in the event at the finish (Figure F-7). There are a variety of ways of displaying the results, from blackboard to ladder-like to a clothesline-type device where each orienteer's name, point score, and times are listed.

i. **Clue Description Card.** These cards are prepared with the master maps after the course is set. They contain the descriptive clues for each control point, control code, grid coordinate references, returning time for competitors, removal times for each location, and panic azimuth (Figure F-8). The terminology on these must be identical to that listed in the definition section. These cards and the master maps must be kept confidential until the orienteers start the event.

j. **Scoring.** The cross-country or free event is scored by the orienteer's time alone. All control points must be visited; failure to visit one results in disqualification. In this event, the fastest time wins.

(1) A variation that can be introduced for novices is to have a not-later-than return time at the finish and add minutes to the orienteer's final time for minutes late and control points not located.

(2) The score event requires the amassing of as many points as possible within the time limit. Points are deducted for extra time spent on the course, usually one point for each 10 seconds extra.

k. Prizes. A monetary prize is not awarded. A sug-gested prize for beginners is an orienteering compass or some other practical outdoor-sports item

SCORE ORIENTEERING EVENT			
NAME _____	COMPANY _____		
NAME _____	COMPANY _____		
TEAM _____			
STARTING TIME _____			
FINISHING TIME _____			
1	6	11	16
2	7	12	17
3	8	13	18
4	9	14	19
5	10	15	20

Total Value of Points
Penalty Points
Final Score

LAND NAVIGATION II

SCORE ORIENTEERING
Description Clues of Control Signs

POINTS	VALUE	DESCRIPTION CLUE
1	10	SE SLOPE OF HILL 1211
2	10	CREST OF HILL 1211
3	10	SADDLE
4	10	EDGE OF RIDGE
5	10	NW CREST OF RIDGE
6	20	CREST OF HILL
7	20	CREST OF HILL
8	20	TOP OF SEVERE DRAW
9	20	SW CREST OF RIDGE
10	10	10 METERS N OF RJ
11	15	CREST OF NW SPUR OF RIDGE
12	10	HIGH POINT OF KNOLL
13	20	HIGH POINT ON HILL SIDE
14	20	NE CORNER OF BLACK CAP MOUNTAINS
15	20	CREST OF SMALL HILL
16	15	NW CREST OF RIDGE
17	10	50 METERS OFF RJ
18	10	10 METERS OFF ROAD
19	10	20 METERS SE OF RJ
20	10	CENTER OF RJ

NOTE: All control signs are located at eye level on trees.

1. ALL WORK IS INDIVIDUAL TEAM EFFORT.
2. YOU MUST NOT JOIN WITH OR COORDINATE WITH ANY OTHER TEAM.
3. YOU MUST PERSONALLY VISIT EACH POINT YOU INDICATE ON YOUR SCORECARD.

Figure F-8. Clue description card.

F-7. SAFETY ON THE COURSE

A first-aid kit must he available at the start and finish. One of the officials should be trained in first aid or have a medic at the event. Other safety measures include:

a. **Control Points.** Locate the controls where the safety of the competitor is not jeopardized by hazardous terrain or other circumstances.

b. **Safety Lane.** Have a location, usually linear, on the course where the competitor may go if injured, fatigued, or lost. A good course will usually have its boundary as a safety lane. Then a competitor can set a panic azimuth on the compass and follow it until he reaches the boundary.

c. **Finish Time.** All orienteering events must have a final return time.

At this time, all competitors must report to the finish line even if they have not completed the course.

d. **Search-and-Rescue Procedures.** If all competitors have not returned by the end of the competition, the officials should drive along the boundaries of the course to pick up the missing orienteers.

F-8. CONTROL POINT GUIDELINES

When the control point is marked on the map as well as on the ground, the description of that point is prefaced by the definite article **the**; for example, **the pond.** When the control point is marked on the ground but is not shown on the map, then the description of the point is prefaced by the indefinite article **a**; for example, **a trail junction.** In this case, care must be taken to ensure that no similar control exists within at least 25 meters. If it does, then either the control must not be used or it must be specified by a directional note in parentheses; for example, a depression (northern). Other guidelines include:

a. Points of the compass are denoted by capital letters; for example, S, E, SE.

b. Control points within 100 meters of each other or different courses are not to be on the same features or on features of the same description or similar character.

c. For large (up to 75 meters across) features or features that are not possible to see across, the position of the control marker on the control point should be given in the instructions. For example, the east side of the pond; the north side of the building.

d. If a very large (100 to 200 meters) feature is used, the control marker should be visible from most directions from at least 25 meters.

e. If a control point is near but not on a conspicuous feature, this fact and the location of the marker should be clearly given; for example, 10 meters E of the junction. Avoid this kind of control point.

f. Use trees in control descriptions only if they are prominent and a totally different species from those surrounding. Never use bushes and fauna as control points.

g. Number control points in red on the master map.

h. For cross-country events, join all control points by a red line indicating the course's shape.

F-9. MAP SYMBOLS

The map symbols in Figure F-9 are typical topographic and cultural symbols that can be selected for orienteering control points. The map cutouts have been selected from DMA maps.

1.
ROAD JUNCTION
Where one road joins
another

2.
ROAD INTERSECTION
Where one road crosses
another

3.
CURVE
A bend in the road

Figure F-9. Map symbols

4.
DEAD END
Where a road stops with
no outlets

5.
TRAIL JUNCTION
Where two trails join

6.
TRAIL DEAD END
Where the trail stops

7.
TRAIL CURVE

8.
ROAD AND TRAIL JUNCTION
Where a road intersects with
two trails

Figure F-9. Map symbols (continued).

9.
ROAD AND TRAIL INTERSECTION
Where a road crosses a trail

10.
INTERMITTENT STREAM JUNCTION

11.
STREAM JUNCTION

12.
BEND IN STREAM

13.
POND'S EASTERNMOST POINT

Figure F-9. Map symbols (continued).

Figure F-9. Map symbol (continued).

19.
ISLAND
Land surrounded by water

20.
HILLTOP
The highest elevation of
a landmark

21.
RIDGE
The trail follows the path of
a ridge

22.
SADDLE
The low point between two or
more points of higher ground

23.
STEEP SIDE OF A HILL

Figure F-9. Map symbols(continued).

24.
DRAWS

25.
SPUR

26.
DRAW OR BEGINNING OF
A VALLEY

27.
SEMILEVEL GROUND ALONG THE
PATH OF A RIDGE

28.
JUNCTION OF STREAMS
Two intermittent streams join
another stream

Figure F-9. Map symbols (continued).

29.
VALLEY
In both directions of a creek

30.
CREEK JUNCTION

31.
DRAW WITH INTERMITTENT
STREAM

32.
HILLTOP

33 .
VALLEY

Figure F-9. Map symbols (continued).

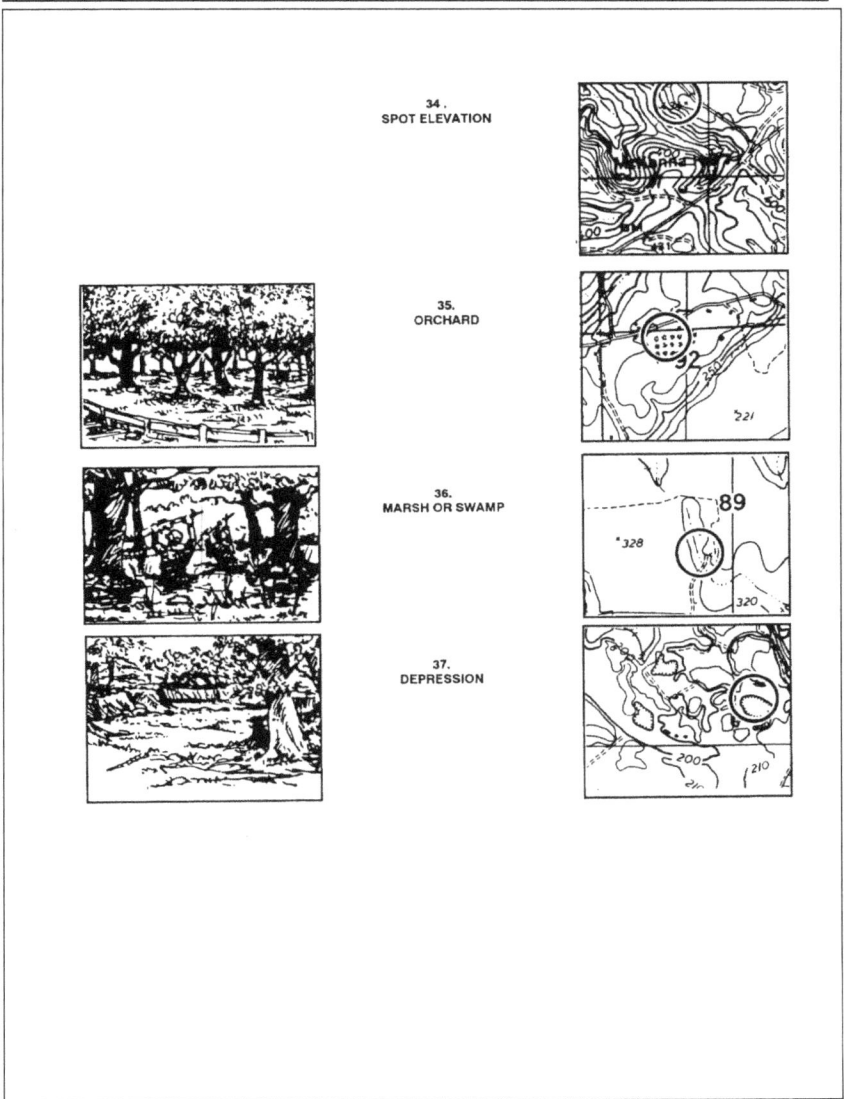

34.
SPOT ELEVATION

35.
ORCHARD

36.
MARSH OR SWAMP

37.
DEPRESSION

Figure F-9. Map symbols (continued).

Figure F-9. Map symbols (continued).

F-10. ORIENTEERING TECHNIQUES

The orienteer should try not to use the compass to orient he map. The terrain association technique is recommended instead. The orienteer should learn the following techniques:

a. **Pacing.** One of the basic skills that the orienteer should develop early is how to keep track of distance traveled while walking and running. This is done on a 100-meter pace course.

b. **Thumbing.** This technique is very sample, but the map has to be folded small to use it. The orienteer finds his location on the map and places his thumb directly next to it. He moves from point to point on the ground without moving his thumb from his initial location. To find the new location, the only thing that he has to do is look at the map and use his thumb as a point of reference for his last location. This technique prevents the orienteer from looking all over the map for his location.

c. **Handrails.** This technique enables the orienteer to move rapidly on the ground by using existing linear features (such as trails, fences, roads, and streams) that are plotted along his route. They can also be used as limits or boundaries between control points (Figure F-10).

d. **Attack Points.** These are permanent known landmarks that are easily identified on the ground. They can be used as points of reference to find control points located in the woods. Some examples of attack points are stream junctions, bridges, and road intersections.

F-11. CIVILIAN ORIENTEERING

Civilian orienteering is conducted under the guidelines of the United States Orienteering Federation with at least 70 clubs currently affiliated. Although civilian orienteering is a form of land navigation, the terms, symbols, and techniques are different from the military.

a. An expert military map reader/land navigator is by no means ready to compete in a civilian orienteering event. However, military experience in navigating on the ground and reading maps will help individuals to become good orienteers. Several orienteering practices and complete familiarization with the map symbols and terms before participating in a real orienteering event is recommended.

(1) *The map*. The standard orienteering map is a very detailed, 1:15,000-scale, colored topographical map. All orienteering maps contain only north-south lines that are magnetically drawn; this eliminates any declination conversions. Because of the absence of horizontal lines, grid coordinates cannot be plotted and therefore are not needed.

(2) *Symbols (legend)*. Despite standard orienteering symbols, the legend in orienteering maps has a tendency to change from map to map. A simple way to overcome this problem is to get familiar with the legend every time that a different map is used.

(3) *Scale*. The scale of orienteering maps is 1:15,000. This requires an immediate adjustment for the military land navigator, especially while moving from

Figure F-10. Handrails.

point to point. It takes a while for a person that commonly uses a 1:50,000 scale to get used to the orienteering map.

(4) *Contours*. The normal contour interval in an orienteering map is 5 meters. This interval, combined with the scale, makes the orienteering maps so meticu-lously detailed that a 1-meter boulder, a 3-meter shallow ditch, or a 1-meter depression will show on the map. This may initially shock a new orienteer.

(5) *Terms and description of clues*. The names of

landforms are different from those commonly known to the military. For example, a valley or a draw is known as a reentrant; an intermittent stream is known as a dry ditch. These terms, with a description of clues indicating the position and location of the control points, are used instead of grid coordinates.

b. The characteristics of the map, the absence of grid coordinates, the description of clues, and the methods used in finding the control points are what make civilian orienteering different from military land navigation.

M2 COMPASS

The M2 compass (Figure G-1) is a rust proof and dust proof magnetic instrument that provides slope, angle of site, and azimuth readings. One of the most important features of the M2 compass is that it is graduated in mils and does not require a conversion from degrees to mils as does the M1 compass. It can be calibrated to provide a grid azimuth or it can be used uncalibrated to determine a magnetic azimuth.

G-1. MAGNETIC NEEDLE

Except for the magnetic needle and its pivot, the compass is made of nonmagnetic materials. When the cover is closed, the magnetic needle is automatically lifted from its pivot and held firmly against the glass window. When the compass is open and leveled, the needle floats freely upon its pivot and points to magnetic north. Note that both ends of the needle are shaped like an arrow, and that one arrow is painted white and the other is black. It is the white end of the needle that points to magnetic north. Because the needle is magnetic, it will also be attracted to large iron or steel objects in the near vicinity, to electrical power lines, and to operating generators (see paragraph 9-3b). Magnetic compass readings measured near such objects are apt to be in error due to the magnetic attraction of these objects.

G-2. CIRCULAR LEVEL

The M2 compass has a circular level that is used to level the instrument when measuring azimuths. The circular level bubble must be centered before reading the azimuth. The compass is equipped with front and rear sights for aligning on the object to which the azimuth is desired.

G-3. COMPASS AZIMUTH SCALE

The compass azimuth scale is a circle divided into 6400 mils. Beginning with zero, the graduations are numbered every 200 mils. The long, unnumbered graduations appearing halfway between the numbered graduations are the odd-numbered hundreds (100, 300, 500, and so forth). Short graduation marks divide each 100-mil segment into equal portions of 20 mils.

a. **Reading the Azimuth Scale.** Azimuths are read
n the azimuth scale from the black end of the compass
dle.

b. **Setting Up the Compass.** To set up the M2
npass, open the cover and fold the rear sight holder out
allel with the face of the compass. Fold the rear sight
perpendicular with its holder. Fold the front sight up,
allel with the mirror. Then fold the cover (mirror)
ard the compass until it is at an angle of approximately
degrees to the face of the compass so that, with your
behind the rear sight, the black end of the compass
dle can be readily viewed in the mirror. The compass
ow set up for measuring an azimuth.

c. **Measuring an Azimuth.** Once the compass is set
up and all steel objects are at least 18 meters away from
your position, you are ready to measure an azimuth. Hold
the compass in both hands at eye level with your arms
braced against your body and with the rear sight nearest
your eyes. Sight through the rear sight and the window in
the mirror and align the hairline at the reflection of the
face of the compass. Center the circular level bubble. With
the bubble centered and the hairline aligned on the object,
look at the mirror reflection of the compass scale and read
the azimuth to which the black end of the needle is
pointing. Remember, magnetic attractions or movement
by you may cause errors in your readings.

ADDITIONAL AIDS

This appendix provides information on the operation and function of already fielded, and soon to be fielded, devices that can be used as aids to navigation.

H-1. AN/PVS-5, NIGHT VISION GOGGLES

These goggles are passive night vision devices. An infrared light source and positive control switch permit close in viewing under limited illumination. The AN/PVS-5 has a field of view of 40 degrees and a range of 150 meters.

a. The device has the capability for continuous passive operation over a 15-hour period without battery replacement. It weighs 1.5 pounds and is face-mounted. An eyepiece diopter is provided so the device can be worn without corrective lenses.

b. It is designed to assist the following tasks: command and control, fire control, reconnaissance, close-in surveillance, terrain navigation, first aid, operation and maintenance of vehicles, selection of positions, traffic control, rear and critical area security, patrolling, combat engineer tasks, radar team employment, resupply activities, and flight-line functions.

c. It is a fielded system used by combat, CS, and CSS elements. The infantry, armor, air defense, field artillery, aviation, engineer, intelligence, military police, transportation, signal, quartermaster, chemical, maintenance, missile, and munitions units all use the device to help accomplish their missions.

d. The AN/PVS-5 can assist the land navigator under limited visibility conditions. Chemical lights may be placed at selected intervals along the unit's route of movement, and they can be observed through the AN/PVS-5. Another navigation technique is to have one person reading the map while another person reads the terrain, both using AN/PVS-5's. This allows the map reader and the terrain interpreter to exchange information on what terrain is observed, both on the map and on the ground. It allows each user to concentrate his AN/PVS-5 on one task. Land navigation, especially mounted, is a task better performed by more than one person. The above technique allows one soldier to perform map interpretation in the cargo portion of the vehicle while another soldier, possibly the driver, tra mits to him information pertaining to the terrain served on the ground.

H-2. AN/PVS-7, NIGHT VISION GOGGLES

The AN/PVS-7 is a lightweight (1.5 pounds), ima intensification, passive night-vision device that uses a bient light conditions. It has the same applications as AN/PVS-5. It is designed to be used in the same way and by the same units as, the AN/PVS-5. The AN/PV has a field of view of 40 meters and a range of 3 meters in moonlight and 150 meters in starlight.

H-3. ENHANCED PLRS USER UNIT

The enhanced position location reporting syst (EPLRS)/joint tactical information distribution syst (JTIDS), hybrid (PJH), is a computer-based system. provides near real-time, secure data communicatio identification, navigation, position location, and au matic reporting to support the need of commanders information on the location, identification, and mo ment of friendly forces.

a. The EPLRS is based on synchronized ra transmissions in a network of users controlled by a m ter station. The major elements of a EPLRS commu include the airborne, surface vehicular, and man-pa users; the EPLRS master station; and an alternate mas station. The system can handle 370 user units in division-size deployment per master station with a ty cal location accuracy at 15 meters. The man-pack u weighs 23 pounds and includes the basic user unit, u readout, antenna, backpack, and two batteries.

b. The EPLRS will be deployed at battalion a company level. Its use allows—

(1) Infantry or tank platoons to locate their po tions, know the location of their friendly units, naviga to predetermined locations, and be informed when ne or crossing boundaries.

(2) Artillery batteries to locate forward observers d friendly units, and position firing batteries.

(3) Aircraft to locate their exact positions; know the ation of other friendly units; navigate to any friendly its, or a location entered by pilot; navigate in selected ght corridors; and be alerted when entering or leaving rridors or boundaries.

(4) Command and control elements at all echelons locate and control friendly units/aircraft.

c. The network control station will be located at gade level to provide position location/navigation and ntification services. It will also provide interface tween the battalion and company systems, and the IDS terminals.

d. It will be fielded to infantry, armor, field artillery, litary police, engineer, intelligence, aviation, signal, d air defense artillery units.

e. The EPLRS is a system that allows units to navi- te from one point to another with the capability of cating itself and other friendly units equipped with the ne system.

4. GLOBAL POSITIONING SYSTEM
e GPS is a space-based, radio-positioning navigation stem that provides accurate passive position, speed, stance, and bearing of other locations to suitably uipped users.

a. The system assists the user in performing such ssions as siting, surveying, tactical reconnaissance, nsor emplacement, artillery forward observing, close support, general navigation, mechanized maneuver, gineer surveying, amphibious operations, signal intel- ence operations, electronic warfare operations, and ound-based forward air control.

b. It can be operated in all weather, day or night, ywhere in the world; it may also be used during nu- ear, biological, and chemical warfare.

c. It has been fielded to infantry, armor, field tillery, military police, engineer, military intelligence, gnal, air defense artillery, and aviation units. (See ppendix J for more information on GPS.)

5. POSITION AND AZIMUTH DETERMINING SYSTEM
e PADS is a highly mobile, self-contained, passive, -weather, survey-accurate position/navigation instru- ent used by field artillery and air defense artillery units r fire support missions. Its basis of issue is two sets per tillery battalion. The device is about the size of a kilowatt generator and weighs 322.8 pounds in opera- onal configuration.

a. The two-man PADS survey party uses the high-

mobility multipurpose wheeled vehicle, the commercial utility cargo vehicle, the small-unit support vehicle, or the M151 1/4-ton utility truck. The system can be transferred while operating into the light observation helicopter (OH-58A) or driven into the CH-47 medium cargo helicopter.

b. The system provides real-time, three-dimensional coordinates in meters and a grid azimuth in mils. It will also give direction and altitude.

c. The PADS can be used by the land navigator to assist in giving accurate azimuth and distance between locations. A unit requiring accurate information as to its present location can also use PADS to get it. The PADS, if used properly, can assist many units in the performance of their mission.

> **WARNING**
> Laser devices are potentially dangerous. Their rays can and will burn someone's eyes If they look directly at them. Users should not direct the beams at friendly positions or where they could reflect off shiny surfaces into friendly positions. Other soldiers must know where lasers are being used and take care not to look directly at the laser beam.

H-6. GROUND-VEHICULAR LASER LOCATOR DESIGNATOR
The G/VLLD is the Army's long-range designator for precision-guided semi-active laser weapons. It is two-man portable for short distances and can be mounted on the M113AI interim FIST vehicle when it has the vehicle adapter assembly. The G/VLLD provides accurate observer-to-target distance, vertical angle, and azimuth data to the operator. All three items of information are visible in the operator's eyepiece display.

a. The G/VLLD is equipped with an AN1AS-4 night sight. This night sight increases the operator's ability to detect and engage targets during reduced visibility caused by darkness or battlefield obscuration.

b. The G/VLLD can give the navigator accurate line-of-sight distance to an object. The system can be used to determine its present location using resection and can assist the navigator in determining azimuth and distance to his objective.

H-7. QUICK RESPONSE MULTICOLOR PRINTER
The QRMP is a self-contained, laser, xerography printer capable of reproducing maps, photographs, annotated graphics, transparent originals, and digital terrain data

in full color on transparent material or standard map paper. The QRMP system will consist of a QRMP housed in an 8 by 8 by 20 feet ISO shelter mounted on a 5-ton truck with a dedicated military-standard 30-kilo-watt generator. Each system will carry at least a seven-day supply of all necessary materials.

a. The QRMP system has map size (24-by 30-inch paper size and 22.5- by 29-inch image size), color printing, scanning and electronics subsystems. It produces the first copy in less than five minutes in full color and sustains a copy rate of 50 to 100 copies per hour for full color products. The system uses a charged couple device array for scanning and sophisticated electronic signal process to electrostatically discharge a selenium photorece drum.

b. The QRMP has the capability to print terrain other graphics directly from digital output from digital topographic support system or another QRM The first unit is scheduled to be equipped with the QR in 1QFY97, and the initial operating capability scheduled for 4QFY97. The QRMP system will be by engineer topographers at division, corps, and echel above corps.

FOREIGN MAPS

The use of foreign maps poses several problems to the land navigator. These products are often inferior in both content reliability and topographic accuracy to those produced by the DMA. Clues to these weaknesses are the apparent crudeness of the maps, unusually old compilation dates, or differences in mapped and actual terrain. The following characteristics should be examined closely.

I-1. HYDROGRAPHY

Of all the symbols on foreign maps, those for hydrography conform most closely to DMA usage. The use of blue lines and areas to depict streams, rivers, lakes, and seas seems to be universally accepted. The one caution to be observed is that foreign cartographers use different sets of rules to govern what is and what is not included on the map. Distinction between perennial and intermittent streams is usually not made.

I-2. VEGETATION

The classification and symbols for vegetation on most foreign maps are different to those used on DMA maps. The vegetation included on many foreign maps is often extensive, identifying not only vegetated areas—but also the specific types of vegetation present. Green is the predominant color used to represent vegetation; but, blue and black are sometimes used. The symbols that depict the various types of vegetation differ greatly from one foreign map to another.

I-3. CULTURAL AND LINEAR FEATURES

Perhaps the most striking difference between DMA and foreign maps is the set of symbols used to portray cultural features. Some symbols found on foreign maps are very unusual. Symbols for linear features on foreign maps are also likely to confuse the user who is accustomed to DMA symbols. DMA uses 10 basic road symbols to portray different classes of roads and trails; foreign mappers use many more.

I-4. TERRAIN RELIEF

Foreign maps generally use contour lines to portray terrain relief, but substantial variability exists in the contour intervals employed. They may range from 5 to 100 meters.

I-5. SCALE

Scales found on foreign maps include 1:25,000, 1:63,360, 1:63,600, 1:75,000, and 1:100,000. Most foreign large-scale topographic maps have been overprinted with 1,000-meter grid squares; so, it is unlikely that the variable scales will have much effect on your ability to use them. However, you must learn to estimate grid coordinates because your 1:25,000 and 1:50,000 grid coordinate scales may not work.

I-6. STEPS TO INTERPRETING FOREIGN MAPS

After discussing the many difficulties and limited advantages encountered when using foreign maps, it is only appropriate that some strategy be offered to help you with the task.

a. In the August 1942 issue of The Military Engineer, Robert B. Rigg, Lieutenant, Cavalry, suggested a five-step process for reading and interpreting foreign maps. It is as appropriate today as it was when he first proposed it.

Step 1. Look for the date of the map first. There are generally four dates: survey and compilation, publication, printing and reprinting, and revision. The date of the survey and compilation is most important. A conspicuous date of revision generally means that the entire map was not redrawn—only spot revisions were made.

Step 2. Note whether the publisher is military, government, or civilian. Maps published by the government or the military are generally most accurate.

Step 3. Look at the composition. To a great extent, this will reveal the map's accuracy. Was care taken in the cartography? Are symbols and labels properly placed? Is the draftsmanship precise? Is the coastline or river bank detailed?

Step 4. Observe the map's color. Does it enhance your understanding of does it obscure and confuse? The importance of one subject (coloring) must warrant cancelling others. If it confuses, the map is probably not very accurate.

Step 5. Begin to decode the various map colors, symbols, and terms. Study these items by examining one feature classification at a time (culture,hydrography, topography, and vegetation). As an accomplished navigator, you should already have a good understanding of your area of operations, so translation of the map's symbols should not present an impossible task. Use your notebook to develop an English version of the legend or create a new legend of your own.

b. In dealing with the challenge of using a foreign map, be certain to use these five steps. In doing so, you are also encouraged to bring to bear all that you know about the geographic area and your skills in terrain analysis, map reading, map interpretation, and problem solving. After careful and confident analysis, you will find that what you do know about the foreign map is more than what you do not know about it. The secret often lies in the fact that the world portrayed on a map represents a kind of international language of its own, which allows you to easily determine the map's accuracy and to decode its colors, symbols, and labels.

GLOBAL POSITIONING SYSTEM

The ability to accurately determine position location has always been a major problem for soldiers. However, the global positioning system has solved that problem. Soldiers will now be able to determine their position accurately to within 10 meters.

J-1. DEFINITION

The GPS is a satellite-based, radio navigational system. It consists of a constellation with 21 active satellites and 3-spare satellites that interfaces with a ground-, air-, or sea-based receiver. Each satellite transmits data that enables the GPS receiver to provide precise position and time to the user. The GPS receivers come in several configurations, hand-held, vehicular-mounted, aircraft-mounted, and watercraft-mounted.

J-2. OPERATION

The GPS is based on satellite ranging. It figures the users position on earth by measuring the distance from a group of satellites in space to the users location. For accurate data, the receiver must track four or more satellites. Three satellites can be used if the user manually inputs the altitude for that location. Most GPS receivers will provide the user with the number of satellites that it is tracking, and whether or not the signals are good. Some receivers can he manually switched to track only three satellites if the user knows his altitude. This method provides the user with accurate data much faster than that provided by tracking four or more satellites. Each type receiver has a number of mode keys that have a variety of functions. To better understand how the GPS receiver operates, refer to the operators' manual.

J-3. CAPABILITIES

The GPS provides worldwide, 24-hour, all-weather, day or night coverage when the satellite constellation is complete. The GPS can locate the position of the user accurately to within 10 meters. It can determine the distance and direction from the user to a programmed location or the distance between two programmed locations called waypoints. It provides exact date and time for the time zone in which the user is located. The data supplied by the GPS is helpful in performing several techniques, procedures, and missions that require soldiers to know their exact location. Some examples are:

· • Sighting.
· • Surveying.
· • Sensor or minefield emplacement.

· • Forward observing.
· • Close air support.
· • Route planning and execution.
· • Amphibious operations.
· • Artillery and mortar emplacement.
· • Fire support planning.

J-4. LIMITATIONS

Until the 21-satellite constellation is complete, coverage may be limited to specific hours of each day in certain areas of the world. The GPS navigational signals are similar to light rays, so anything that blocks light will reduce or block the effectiveness of the signals. The more unobstructed the view of the sly, the better the system performs.

J-5. COMPATABILITY

All GPS receivers have primarily the same function, but the input and control keys vary between the different receivers. The GPS can reference and format position coordinates in any of the following systems:

- **Degrees, Minutes, Seconds** (DMS): Latitude/longitude-based system with position expressed in degrees, minutes, and seconds.
- **Degrees, Minutes** (DM): Latitude/longitude-based system with position expressed in degrees and minutes.
- **Universal Traverse Mercator (UTM):** Grid zone system with the northing and easting position expressed in meters.
- **Military Grid Reference System** (MGRS): Grid zone/grid square system with coordinates of position expressed in meters.

The following is a list of land navigation subjects from other sections of this manual in which GPS can be used to assist soldiers in navigating and map reading:

a. **Grid Coordinates (Chapter 4)**. GPS makes determining a 4-, 6-, 8-, and 10-digit grid coordinate of a location easy. On most GPS receivers, the position mode will give the user a 10-digit grid coordinate to their present location.

b. **Distance (Chapter 5) and Direction (Chapter 6**). The mode for determining distance and direction depends on the GPS receiver being used. One thing the different types of receivers have in common is that to determine direction and distance, the user must enter at least one waypoint (WPT). When the receiver measures direction and distance from the present location or from waypoint to waypoint, the distance is measured in straight line only. Distance can be measured in miles, yards, feet, kilometers, meters, or nautical knots or feet. For determining direction, the user can select degrees, mils, or reds. Depending on the receiver, the user can select true north, magnetic north, or grid north.

c. **Navigational Equipment and Methods (Chapter 9).** Unlike the compass, the GPS receiver when set on navigation mode (NAV) will guide the user to a selected

waypoint by actually telling the user how far left or right the user has drifted from the desired azimuth. With this

option, the user can take the most expeditious route possible, moving around an obstacle or area with out replotting and reorienting.

d. **Mounted Land Navigation (Chapter 12).** While in the NAV mode, the user can navigate to a waypoint using steering and distance, and the receiver will tell the user how far he has yet to travel, and at the current speed, how long it will take to get to the waypoint.

e. **Navigation in Different Types of Terrain (Chapter 13).** The GPS is capable of being used in any terrain, especially more open terrain like the desert.

f. **Unit Sustainment (Chapter 14).** The GPS can be used to read coordinates to quickly and accurately establish and verify land navigation courses.

GLOSSARY

ACRONYMS AND ABBREVIATIONS

AA .... avenue of approach
ANCOC......... advanced noncommissioned
　　　　　　　officer course
AR .... Army regulation
BNCO............ Cbasic noncomissioned officer course
BT .... basic training
cm.................. centimeter
CONUS .... continental United States
CS.................. combat support
CSS............... combat service support
CUCV............ commercial utility cargo vehicle
DD .... Department of Defense
DMA.............. Defense Mapping Agency
E .... east
EPLRS .... enhanced position location
　　　　　　　reporting system
FIST .... fire support team
FM................. field manual
FORSCOM... United States Army
　　　　　　　Forces Command
GD................. ground distance
GEOREF....... geographic reference system
G-M............... grid-magnetic
GPS .... global positioning system
GSR............... ground surveillance radar
GTA .... graphic training aid
G/VLLD........ ground/vehicular laser
　　　　　　　locator designator
HD................. horizontal distance
HHC.............. headquarters and headquarters
　　　　　　　company
HMMWV...... high-mobility multipurpose
　　　　　　　wheeled vehicle
JOG............... joint operations graphics
JTIDS............ joint tactical information
　　　　　　　distribution system
km .... kilometer
LAT............... latitude
MD .... map distance

METT-T........ mission, enemy, terrain, troops
　　　　　　　and time available
MITAC.......... map interpretation and
　　　　　　　terrain association course
N .... north
NCO............... noncommissioned officer
OAC............... officer advanced course
OBC.............. officer basic course
OCS .... officer candidate school
OSUT............ one station unit training
PADS............. position and azimuth
　　　　　　　determining system
PD................. photo distance
PJH............... hybrid (PLRS and JTIDS)
PLDC............ primary leadership development course
POI .... program of instruction
PRE .... precommission
QRMP .... quick response multicolor printer
RF .... representative fraction
ROTC........... Reserve Officers' Training Corps
S..................... south
SF.................. standard form
SOSES............ shapes, orientations, sizes, elevations,
　　　　　　　and slopes
SUSV .... small-unit support vehicle
tan.................. tangent
TM................. technical manual
TOW.............. tube-launched, optically tracked,
　　　　　　　wire-guided missile
TRADOC....... Training and Doctrine Command
topo. .... topographic
UPS............... universal polar stereographic
US .... United States
USGS............ United States Geological Survey
UTM.............. universal transverse mercator
VO .... vertical distance
VNAS............ vehicular navigation aids system
W .... west

REFERENCES

SOURCES USED
These are the sources quoted or paraphrased in this publication.

FM 21-31.	**Topographic Symbols,** 19 June 1961.
FM 25-4.	**How to Conduct Training Exercises,** 10 September 1984.
FM 25-100.	**Training the Force,** 15 November 1988.
FM 25-101.	**Battle Focused Training,** 30 September 1990.
FM 101-5-1.	**Operational Terms and Symbols,** 21 October 1985.

DOCUMENTS NEEDED
These documents must be available to the intended users of this publication.

AR 115-11.	**Army Topography,** 01 March 1980.
AR 380-5.	**Department of the Army Information Security Program,** 25 February 1988.
AR 380-40.	**Policy for Safeguarding and Controlling Communication and Security (COMSEC) Materal,** 22 October 1990.
FM 5-33.	**Terrain Analysis,** 11 July 1990.
FM 34-1.	**Intelligence and Electronic Warfare Operations,** 02 July 1987.
FM 34-3.	**Intelligence Analysis,** 15 March 1990.
FM 101-10-1.	**Staff Officers Field Manual: Organizational, Technical, and Logistical Data,** 07 October 1987.
TC 6-40.	**Field Artillery Manual Cannon Gunnery,** 27 December 1988.
TM 5-240.	**Compilation and Color Separation of Topographic Maps,** 15 June 1971.
TM 5-243.	**Cartographic Aerial Photography,** 02 January 1970.

INDEX

By Order of the Secretary of the Army:

GORDON R. SULLIVAN
General, United States Army
Chief of Staff

Official:

Milton H. Hamilton

MILTON H. HAMILTON
Administrative Assistant to the
Secretary of the Army
03794

DISTRIBUTION:

Active Army, USAR, and ARNG: To be distributed in accordance with DA Form 12-11E, requirements for FM 21-26, Map Reading and Land Navigation (Qty rqr block no. 0166).

U.S. GOVERNMENT PRINTING
OFFICE:1994-300-421/02100